The Art and Science of Coffee Fermentation

Marcel Hackler • Vinzenz Särchen

The Art and Science of Coffee Fermentation

A Guide to Biotransformation

Marcel Hackler
Käthe Kaffeerösterei GmbH
Oldenburg, Germany

Vinzenz Särchen
Rostock, Germany

ISBN 978-3-031-91598-7   ISBN 978-3-031-91599-4   (eBook)
https://doi.org/10.1007/978-3-031-91599-4

© The Editor(s) (if applicable) and The Author(s), under exclusive license to Springer Nature Switzerland AG 2025

This work is subject to copyright. All rights are solely and exclusively licensed by the Publisher, whether the whole or part of the material is concerned, specifically the rights of translation, reprinting, reuse of illustrations, recitation, broadcasting, reproduction on microfilms or in any other physical way, and transmission or information storage and retrieval, electronic adaptation, computer software, or by similar or dissimilar methodology now known or hereafter developed.
The use of general descriptive names, registered names, trademarks, service marks, etc. in this publication does not imply, even in the absence of a specific statement, that such names are exempt from the relevant protective laws and regulations and therefore free for general use.
The publisher, the authors and the editors are safe to assume that the advice and information in this book are believed to be true and accurate at the date of publication. Neither the publisher nor the authors or the editors give a warranty, expressed or implied, with respect to the material contained herein or for any errors or omissions that may have been made. The publisher remains neutral with regard to jurisdictional claims in published maps and institutional affiliations.

Cover illustration: This picture was taken by Philipp Tresbach 2024 in Peru during a farm visit of the cooperative Rutas del Inca. It shows Marcel picking ripe coffee cherries.

This Springer imprint is published by the registered company Springer Nature Switzerland AG
The registered company address is: Gewerbestrasse 11, 6330 Cham, Switzerland

If disposing of this product, please recycle the paper.

CONTENTS

List of Abbreviations	vii
1 INTRODUCTION	1
2 COFFEE FLAVOR DEVELOPMENT	**5**
Coffee Cherry	9
Aromatic Building Blocks of the Coffee Cherry	11
3 BASICS OF FERMENTATION	**15**
Biotechnological Definition of Fermentation	16
What Does the Fermentation of Coffee Need?	18
Environmental Influencing Factors in the Fermentation Process	20
Biotechnologically Utilized Fermentation Types	24
Commonly Used Terms for "Fermentation" in the Coffee Scene	33
4 THE MAIN ACTORS	**41**
Fungi	43
Bacteria	46
Enzymes	50
5 COFFEE PROCESSING	**57**
Why are There so Many Different Methods?	60
Natural/Dry-Processed/Sun-Dried	61
Washed	70
Pulped Natural/Honey/Semi-Dry	74
Burundian	77
Semi-Washed/"Giling Basah"/Wet-Hulled	77
Animal Digestive Fermentation	78
Monsooned	79
"Modern Techniques"	79
6 LET'S PUT IT ALL TOGETHER	**85**
Fermentation and its Impact on Sensory Perception	89
7 SUSTAINABILITY & WASTEWATER	**101**
8 SOME LAST WORDS	**107**
Further Thoughts of us	110
Acknowledgement	111

List of Abbreviations

AAB: Acetic Acid Bacteria
ATP: Adenosine Triphosphate
a_W: Water activity
CGA: Chlorogenic Acid
CO_2: Carbon Dioxide
CoA: Coenzym A
GABA: Gamma-Aminobutyric Acid
LAB: Lactic Acid Bacteria
OTA: Ochratoxine A
pH: pH-value
sp.: (Lat. species) refers to a single, unnamed species of a genus
spp.: (Lat. species pluralis) refers to multiple, unnamed, not further specified species of a genus

1

INTRODUCTION

For thousands of years, food production, processing, and preservation have been closely intertwined with microbial fermentation. When you think of fermentation, sauerkraut, yogurt, kimchi, or fresh sourdough bread with kefir butter probably comes to mind, rather than a sweet-balanced coffee.

However, the advent of elaborate, natural, and anaerobic-processed coffees has made "fermentation" a common term, especially in the specialty coffee scene. The precision in using terms and accurately describing fermentation processes often falls short. A wealth of scientific knowledge exists on coffee fermentation, but the scientific jargon makes it unapproachable.

As coffee enthusiasts who had the opportunity to receive training in biotechnology, our motivation drove us to gather expert knowledge on coffee fermentation and make it accessible to the coffee world. Don't worry – you can understand this book without a degree in biology or chemistry.

One thing is clear: Fermentation is not an invention of humans but an essential natural process for gaining energy from the decomposition of organic matter. Without it, "the earth would be a gigantic landfill where the dead pile up, and there would be no food for the living," as Michael Pollan aptly puts it [1]. Humans and some animal species have learned to utilize fermentation processes by controlling and stopping them immediately – controlled fermentation results in nutritious, aromatic products with often intensified taste.

Fermentation
The conversion of organic raw materials, such as wort in brewing beer, mash in winemaking, cabbage for sauerkraut, or aroma-building blocks in coffee, through microorganisms or the addition of crucial enzymes. This conversion can occur both without oxygen (anaerobic) and with the addition of air (aerobic). The fermentation process produces organic acids, alcohols, and gases.

This book provides an overview of the fundamental impact of fermentation on coffee production. We also explore the various processes and diligent microorganisms and their influence on coffee as we know it. We hope to broaden your perspective on fermentation, green coffee processing and their effects on taste after roasting.

We dedicate this book to all the biological helpers who make coffee what it is: delicious.

So let's get nerdy.

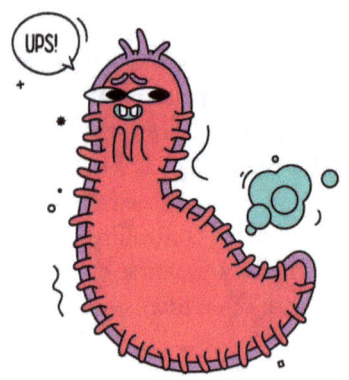

References of Chapter 1

1) Pollan M (2014) **Kochen:** Eine Naturgeschichte der Transformation. Kunstmann A, München

2

COFFEE FLAVOR DEVELOPMENT –
FINGER-LICKING GOOD

Chapter Summary

The relationship between taste and coffee is complex, influenced by a range of factors including the processing, roasting, and preparation methods used, as well as the quality of the coffee itself. Coffee books often explore the aromatic compounds and the flavors that emerge during roasting. However, fermentation plays a significant role in shaping the taste of coffee. While traditional coffee processing methods, like Washed, Honey, and Natural, focus on removing the fruit layers surrounding the coffee cherry, the exact processes of fermentation and their impact on flavor are often underexplored. Fermentation helps remove the mucilage (a gelatinous layer) from the coffee cherry, but its role goes beyond just this physical separation. The fermentation process influences the coffee's flavor profile by interacting with the compounds in the mucilage, which contains e.g. sugars, organic acids, and proteins that contribute to the overall taste.

When delving into the relationship between taste and coffee, one finds that coffee books often discuss the aromatic compounds present in green coffee and the variety of flavors that emerge during the reactions in coffee roasting. The quality of processing, roasting, and preparation can significantly influence the taste of coffee. There is already a good overview of traditional processing methods, from Washed, Honey to Natural. Ultimately, the specialty coffee scene has advanced knowledge regarding the influence of preparation methods (temperature, water quality), quality of coffee (specialty coffee), and the procedure of roasting.

Fig. 2.1: From plant to cup – coffee's lifeline.

Coffee Flavor Development

Maillard Reaction

A chemical reaction or network of chemical reactions. At high temperatures (e.g., roasting, frying, baking), the starting materials, amino compounds, for example react with sugars to produce various aromatic products. Everyone is familiar with the typical roasting aromas developed by frying or baking. However, undesirable aroma compounds can also form, such as acrylamides or bitter substances. These undesirable flavors are specially produced in burnt food – so one should proceed cautiously. The reaction is named after Louis C. Maillard (*1878, †1936), who first reported it in 1912 [1].

Fig. 2.2: Marcel is cupping coffee samples.

Fig. 2.3: Small-batch roaster for coffee samples.

But what about fermentation and taste? As Lee clearly shows, the role of fermentation is in removing the fruit layers that envelop the coffee cherry. Less in focus, however, is the exact processes of fermentation and their impact on the taste and aroma of the coffee itself [2]. And that is astonishing, considering the taste differences of the same coffee with different processing methods in comparison (if you haven't done it yet – try it yourself and get the same coffee from a producer that has been processed differently, e.g., Honey and Natural, and compare them in terms of taste). Nearly 80 years ago, Pederson and Breed (1946) showed the better quality of fermented coffee over non-fermented coffees [3]. Hold up, non-fermented coffee? From a biotechnological perspective, we would immediately argue: Doesn't fermentation always occur when coffee is processed, matures, and dries? So, isn't it instead the different types or intensities of fermentation? Indeed, but more on that later! If you can't wait, take a look at chapter 5.

In traditional coffee processing, fermentation is essential in removing the mucilage, a layer of the coffee cherry. Today, this step plays a unique role in specifically influencing the flavor profiles of coffee and its quality [4].

Since the coffee flavor profile development starts with the coffee cherry, we first want to examine the central element of our beloved favorite drink – the coffee cherry.

Coffee Flavor Development

Fig. 2.4: Coffee seeds in their parchment. Visible is the slimy sticky mucilage and the separated pulp and skin of the depulped cherries.

Coffee Cherry

The coffee cherries serve as a substrate for the microorganisms, acting as the base, a nutrient supplier on which they live. The next illustration shows that the coffee cherry consists of several layers, varying in thickness and consistency. The two coffee seeds (commonly called beans in roasted coffee) are surrounded from inside to out by a silver skin, then the parchment skin, the mucilage, the pulp and finally, the outer skin[5]. The mucilage plays a central role in fermentation. This mucous layer (pectin layer) is a hydrogel made of water, a wide range of different sugar molecules (arabinose, galactose, xylose, rhamnose, and other polysaccharides), lipids and fatty acids (linoleic acid, palmitic acid, oleic acid), organic acids (caffeic acid, quinic acid, oxalic acid), and proteins[4–8]. Since the molecules of interest to microorganisms mainly reside in the mucilage and the pulp, fermentation predominantly occurs there.

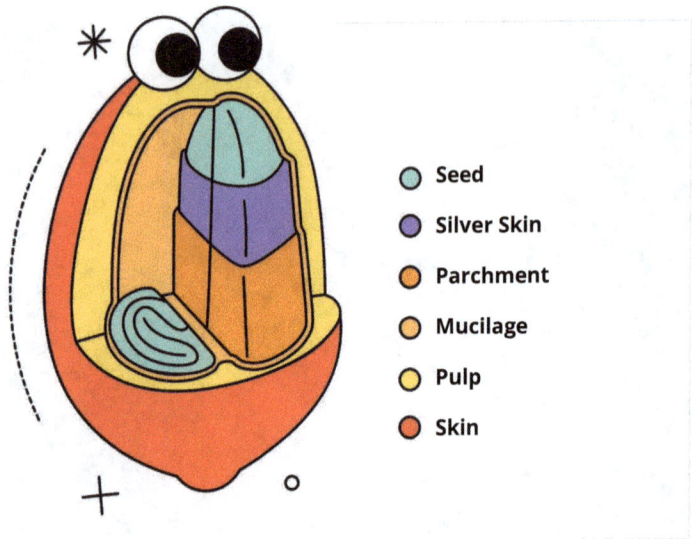

Fig. 2.5: The inner life of a coffee cherry.

Pectin and Hydrogel
Networks of polysaccharides or their corresponding ionic acids (here, galacturonic acid). Those networks can store water without dissolving itself. Thus, it acts as a framework for water storage (by the way, baby diapers work similarly). The combination of a stable framework of biopolymers and the ability to store water is known as a hydrogel. In the coffee cherry, pectin and water form the mucilage between the pulp and the parchment skin.

Coffee Flavor Development

Aromatic Building Blocks of the Coffee Cherry

Let's zoom into the mucilage more closely and examine the aromatic building blocks. These precursors, already present in green coffee, later produce delicious aromas during roasting. For green coffee, the presence of these aromatic building blocks is relevant, and their proportional composition (glucose vs. fructose vs. mannose vs. galactose vs. arabinose) also plays an important role [9]. Here's an example: The sugar composition in raw arabica beans from Tanzania and Mexico are relatively similar, but the raw Mexican beans have an increased total sugar content [10]. With that they are offering more food for the microorganisms. Which results in a different fermentation process along the coffee processing.

In addition to simple and complex sugars (carbohydrates, mono-/poly-saccharides), other organic compounds such as proteins and acids contribute to the sensory properties [11, 12]. The showcase on the next page presents these raw materials. As mentioned above, parts of these raw materials transform during roasting and collectively produce the coffee flavors we all know and love. However, there are also aromatic precursors that, after roasting, create negative qualities and are undesirable, such as butyric acid, but more on that later [13 – 18].

Depending on the processing method, different levels of the following components exist in the same coffee varieties.

Sugar
Sucrose, reducing sugars like fructose, glucose, arabinose, galactose, mannose, rhamnose.

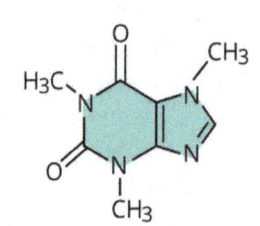

Nitrogen compounds
Such as caffeine, trigonelline, or free amino acids like glutamic acid and aspartic acid.

Organic Acids/Sugar Acids
Such as chlorogenic acid, acetic acid, uronic acid.

Lipids
Including Carotenoids.

Fig. 2.6: Coffee aroma building blocks.

The composition of the raw materials present in green coffee depends on the country and location of cultivation (climatic growth conditions, altitude, region, weather conditions), type of coffee (*C. arabica* vs. *C. robusta*), plant type (varieties like Pacamara or SL 28), degree of ripeness, harvesting time, and the fermentation processes during coffee processing after harvest.

Coffee Flavor Development

References of Chapter 2

1) **Maillard LC (1912)** Action des acides amines sur les sucres; formation des lenadoidines par voie methodique. C R Hebd Seances Acad Sci, Ser. C:66–68
2) **Lee LW, Cheong MW, Curran P, et al (2015)** Coffee fermentation and flavor – An intricate and delicate relationship. Food Chemistry 185:182–191. https://doi.org/10.1016/j.foodchem.2015.03.124
3) **Pederson CS, Breed RS (1946)** Fermentation of coffee. J Food Sci 11(2):99–106. https://doi.org/10.1111/j.1365-2621.1946.tb16331.x
4) **Haile M, Kang WH (2019)** The role of microbes in coffee fermentation and their impact on coffee quality. J Food Qual 2019(1):4836709. https://doi.org/10.1155/2019/4836709
5) **Avallone S, Guiraud JP, Guyot B et al (2000)** Polysaccharide constituents of coffee-bean mucilage. J Food Sci 65(8):1308–1311. https://doi.org/10.1111/j.1365-2621.2000.tb10602.x
6) **Schwan RF, Wheals AE (2003)** Mixed microbial fermentations of chocolate and coffee. In: Boekhout T, Robert V. (ed) Yeasts in Food. Woodhead Publishing Ltd, Cambridge, p 429–449. https://doi.org/10.1533/9781845698485.429
7) **Osorio Pérez V, Álvarez-Barreto CI, Matallana LG et al (2022)** Effect of prolonged fermentations of coffee mucilage with different stages of maturity on the quality and chemical composition of the bean. Fermentation 8(10):519. https://doi.org/10.3390/fermentation8100519
8) **Avallone S, Guiraud JP, Guyot B et al (2001)** Fate of mucilage cell wall polysaccharides during coffee fermentation. J Agric Food Chem 49(11):5556–5559. https://doi.org/10.1021/jf010510s
9) **Mussatto SI, Machado EMS, Martins S et al (2011)** Production, composition, and application of coffee and its industrial residues. Food Bioprocess Technol 4(5):661–672. https://doi.org/10.1007/s11947-011-0565-z

10) Knopp S, Bytof G, Selmar D (2006) Influence of processing on the content of sugars in green Arabica coffee beans.
Eur Food Res Technol 223(2):195–201.
https://doi.org/10.1007/s00217-005-0172-1

11) Franca AS, Mendonça JCF, Oliveira SD (2005) Composition of green and roasted coffees of different cup qualities. LWT 38(7):709–715.
https://doi.org/10.1016/j.lwt.2004.08.014

12) Franca AS, Oliveira LS, Mendonça JCF et al (2005) Physical and chemical attributes of defective crude and roasted coffee beans.
Food Chem 90(1-2):89–94

13) Farah A, Monteiro MC, Calado V et al (2006) Correlation between cup quality and chemical attributes of Brazilian coffee.
Food Chem 98(2):373–380
https://doi.org/10.1016/j.foodchem.2005.07.032

14) Agate AD, Bhat JV (1966) Role of pectinolytic yeasts in the degradation of mucilage layer of Coffea robusta cherries.
Appl Microbiol 14(2):256–260.
https://doi.org/10.1128/am.14.2.256-260.1966

15) Amorim HV, Amorim VL (1977) Coffee enzymes and coffee quality.
In: Robert LO, Allen JSA (ed) Enzymes in food and beverage processing, vol 47. ACS Symposium Series, Washington D.C., p 27–56

16) de Melo Pereira GV, Soccol VT, Pandey A et al (2014) Isolation, selection and evaluation of yeasts for use in fermentation of coffee beans by the wet process. Int J Food Microbiol 188:60–66.
https://doi.org/10.1016/j.ijfoodmicro.2014.07.008

17) de Melo Pereira, GV, Neto E, Soccol VT et al (2015) Conducting starter culture-controlled fermentations of coffee beans during on-farm wet processing: Growth, metabolic analyses and sensorial effects.
Food Res Int 75:348–356.
https://doi.org/10.1016/j.foodres.2015.06.027

18) Evangelista SR, da Cruz Pedroso Miguel MG, Silva CF et al (2015) Microbiological diversity associated with the spontaneous wet method of coffee fermentation. Int J Food Microbiol 210:102–112.
https://doi.org/10.1016/j.ijfoodmicro.2015.06.008

3

BASICS OF FERMENTATION

Chapter Summary

Fermentation is a microbial process that can be industrially utilized. Different fermentation processes rely on various substrates and the resulting intermediates and end products differ. With this, it is possible to significantly influence the aroma. Thus, microbial fermentation is a central adjustment screw for the aromatic properties of coffee, considerably affecting its quality. However, it must also taste good because perhaps you know a situation where you serve your friends an Anaerobic-Natural coffee with shining eyes, and they can't relate to the "funky wine fruitiness" and find it not as tasteful. Therefore, precise knowledge of the underlying processes helps use and control fermentation purposefully so that the sensory quality matches (and nothing goes wrong!). It is interesting to know which processing steps and microorganisms produced the specific aroma that a roaster has teased out and a barista skillfully brews into the cup.

As we have already learned, fermentation plays a crucial role for us humans in the processing of food. In coffee, people traditionally use fermentation to remove the mucilage. However, what exactly is fermentation?

Biotechnological Definition of Fermentation

From a technical or biotechnological perspective, fermentation can generally be described as follows: Fermentation exploits metabolic processes to convert a substrate into one or more desired, plus undesired products. People often use bioreactors for this purpose. These are closed tanks or containers like barrels, with inlets and outlets, where one can control the processes.

Fig. 3.1: A simple bioreactor: Plastic bags are used for fermentation of depulped coffee in Washed or Honey processing by Medim Jimenez in Bellavista, Peru.

Basics of Fermentation

Thus, according to our understanding, fermentation is a microbial process that can be industrially utilized. The wording "fermentation" is interesting because, in English-speaking countries, people often equate fermentation exclusively with anaerobic processes. However, in German-speaking areas anaerobic processes receive the term "Gärung". Therefore, in the German-speaking context, fermentation encompasses both anaerobic (without oxygen, Gärung) and aerobic (with the addition of air) processes.

Aerobic / Anaerobic / Facultative / Microaerophilic / Aerotolerant

The designation originates from the ancient Greek word *aer*, meaning air. It describes the conditions under which an organism metabolizes, whether it metabolizes with oxygen (A), without oxygen (B), whether both conditions are okay (C), whether it prefers oxygen only in low concentration (D) (thus, lives somewhat away from the surface), or whether oxygen does not affect it at all (E)[1].

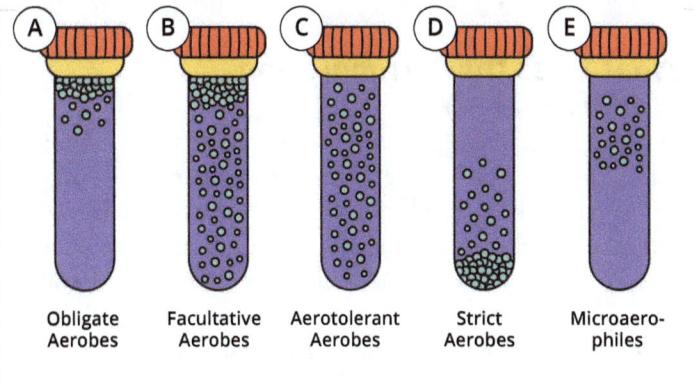

Fig. 3.2: Oxygen dependent living conditions of microorganisms.

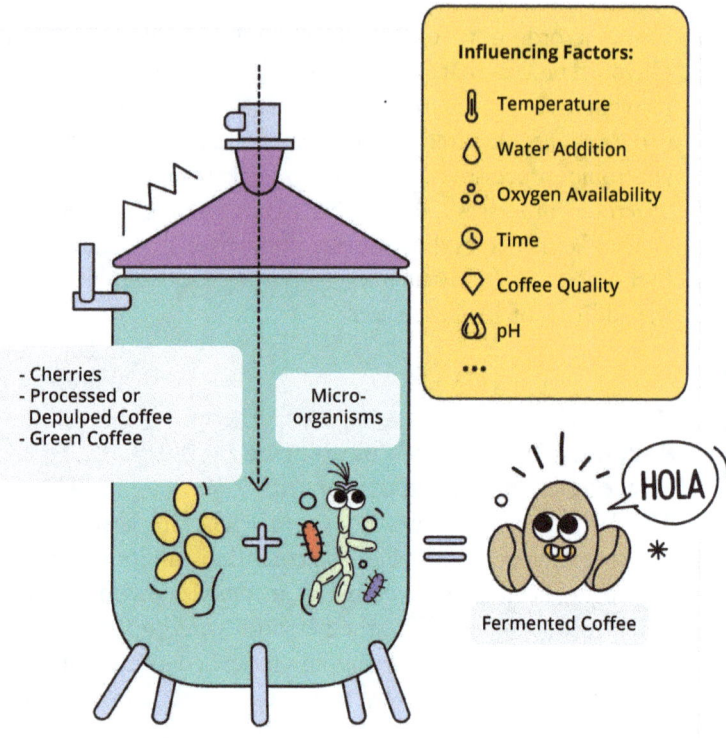

Fig. 3.3: Bioreactor, its components and the factors of influence on coffee fermentation.

What Does the Fermentation of Coffee Need?

In coffee processing, microorganisms are always surrounding coffee cherries. Raw coffee beans can spontaneously trigger fermentation. People call this "wild" fermentation when they add no extra organisms. This process transforms, breaks down, or even initially creates the basic building blocks of coffee aromas. Our environment doesn't just contain a single, isolated microbial species but communities of various microorganisms. When metabolizing the basic building blocks in raw coffee, many processes run in parallel, allowing many different

Basics of Fermentation

intermediate stages of coffee aromas to emerge, depending on microbial diversity and external conditions. The complexity increases as one microbial species' metabolic products can act as substrates for another species' metabolic turnover. Therefore, some aromatic precursors only come into existence through the metabolic activity of our tiny helpers during fermentation.

Microorganisms and their biochemical (metabolic) processes propel the fermentation process forward. Like every living being on our planet, microorganisms possess their metabolism to generate the energy they need. These cellular processes don't happen independently but require special enzymes to facilitate the breakdown (catabolism) and creation (anabolism) of substances into metabolic products. Researchers have identified the enzymes involved in many of these processes and can produce them in isolation. These isolated enzymes bypass the need for organisms in the production and transformation of products. In such scenarios, the specific enzymatic reaction proceeds isolated from the rest of the cell. Fermentation encompasses this cell-free, isolated enzymatic reaction. So, what's this again? Yes, exactly – fermentation can indeed occur without organisms. Crazy, isn't it?

Enzymes
Biological catalysts that can speed up a specific chemical reaction under mild conditions (no extreme temperature and pressure ranges). Biological metabolism, the creation of energy carriers, can only occur with the help of enzymatic proteins.

However, let us look at fermentation step by step, starting with the influencing factors that shape the nature and course of fermentation and thus can also affect aroma and taste.

Environmental Influencing Factors in the Fermentation Process

1. Fermentation Starting Products
Microbial metabolism can alter the content of starting materials that are important for roasting (for Maillard reactions) to produce the desired volatile flavor compounds. Microorganisms thus transform substances, and as they do so, the content of the starting products changes – for example, the remaining free sugar or amino acids [2]. As microorganisms are ubiquitous present, fermentation occurs directly after harvest. Therefore, the aroma profile after roasting is directly affected by the processing [3].

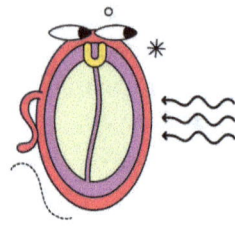

2. Intermediates
The activity of microorganisms produces microbial intermediates (called metabolites). These metabolites, not originating from the coffee plant but purely microbial, can be absorbed by the coffee bean during fermentation and influence the aroma profile [4].

3. Germination
Fermentation is thought to affect the ripening processes of coffee beans. When germination processes start in the coffee beans, this leads to transformations in the chemical composition of the coffee seed itself and, thus, the desired relevant aroma precursors for roasting [4, 5].

Basics of Fermentation

4. Microbial Competition

Naturally occurring microorganisms always live in mixed cultures. Depending on environmental conditions, different organisms are more active than others. Metabolic pathways within the microbial mixed culture can also be coordinated so that the products of one are the required substrates for another microbial species. Unwanted microorganisms, such as mold, can also occur, negatively impacting the coffee aroma and quality. Therefore, paying meticulous attention to a clean working environment and influencing the microbial milieu is crucial [6, 7].

5. Environmental Conditions

Environmental conditions are super important as they influence the growth and presence of various microorganisms. These can also change over time during the fermentation (or at different heights in the fermentation vessel). These include sugar concentration (°Brix), climate, and altitude, which are linked to the temperature variation day and night, time, water activity (a_W), acid-base balance, and oxygen content [2].

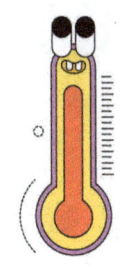

Fig. 3.4: Mountain view 1500 meters above sea level, in the surroundings of San Ignatio, Peru.

6. Water Activity a_w-Value

Water activity describes the amount of water available for microbiological activities. The water content can be higher but bound and not "actively" available. Therefore, the a_w-value is essential for food processing, as different water activities favor different organisms or influence enzymatic and chemical reactions[8]. It is also crucial for storage and transport stability, as from a certain degree of drying and low water activity, the shelf life is increased (e.g. dried fruits compared to fresh fruits)[9].

7. Change in Acid-Base Balance

During fermentation, the acid-base balance within the beans changes. Microorganisms produce organic acids such as lactic or acetic acid, increasing the acidity in the beans. This can, in turn, influence aroma formation during roasting[3,10]. For example, think of anaerobically fermented coffees that can smell very sour in their dried green state and tend to have fermentative and winey characteristics when roasted.

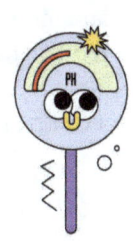

8. Oxygen Availability

The aspect of oxygen availability is significant, as different conditions determine which organisms are active and which metabolic pathways are present. For example, when an oxygen-rich (aerobic) or oxygen-poor (anaerobic) environment exists, Knopp describes how this can lead to naturally fermented beans having a higher glucose and fructose content since less sugar metabolism occurs under aerobic conditions [11]. Thus, more is preserved.

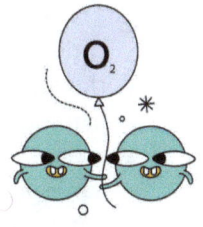

9. Fermentation Time

Time plays a significant role in fermentation. Naturally, microorganisms processing the starting materials in the mucilage during fermentation need time. Therefore, if fermentation ends too early (under-fermentation), it can lead to incomplete metabolism. Suppose fermentation lasts too long (over-fermentation). In that case, the microorganisms have more than enough time to produce unwanted secondary products in sufficient quantities, affecting the sensory quality of the coffee beans.

Here you have it, a first overview of the various adjustments and influencing factors that are particularly important during fermentation. As you can see, there is a pretty wide range. Now, let's look at the different types of fermentation that play a central role in implementing our coffee aroma precursors.

Biotechnologically Utilized Fermentation Types

Fermentation involves converting a starting material under specific conditions (either by microorganisms or cell-free directly with enzymes) into one or more desired products. The individual fermentation types are usually named after their main final product, which is of economic interest. In the following, we will introduce several microbial metabolic pathways. We aim to give you a broad overview of the possible actions in the fermenting coffee cherry, resulting in different flavor profiles in the final coffee in your cup. Due to the sheer number of different metabolism-carrying microorganisms, the list cannot be complete.

Cellular Respiration

Every cell needs energy, whether it originates from bacteria or exists in our bodies. Cells produce this energy through degradative metabolic pathways. "Degradative" means that the cell takes energy-rich substances, breaks them down step by step, and releases energy. However, the released energy does not dissipate as thermal energy, but is transferred to other energy carriers through enzymatic processes.

Basics of Fermentation

ATP

Adenosine Triphosphate represents a biologically universal form of energy. This form is conserved across all organisms, meaning humans produce and use ATP for energy like bacteria, algae, and higher plants. This molecule clearly shows the evolutionary lineage of all Earth's living beings.

Adenosine Triphosphate (ATP) is a critical cellular energy carrier. Energy-rich substrates, such as fats and sugar molecules provide energy. For simplicity, let's focus on sugar molecules, including hexoses (molecules with six carbon atoms, like glucose and fructose) and pentoses (molecules with five carbon atoms, like ribose and xylose). Cells take these sugar molecules and funnel them into the first metabolic pathway, glycolysis. As the name suggests, glycolysis splits the ingested 6-carbon sugar into two smaller 3-carbon sugars, pyruvate, and produces two energy-carrier ATP molecules. This metabolic pathway occurs under both anaerobic and aerobic conditions. It is a general metabolic pathway for many organisms. For the cell, pyruvate is the ultimate energy supplier, as it releases further energy equivalents through another metabolic pathway (tricarboxylic acid cycle and electron transport chain) within the cell's powerhouses (the mitochondria), using oxygen.

25 ATP molecules are produced from two pyruvate molecules, providing energy for the cell's metabolic processes. This process of energy production is essential for us humans, too.

Alcoholic Fermentation

Without oxygen pyruvate can no longer be metabolized through the tricarboxylic acid cycle. As we have learned above, pyruvate is the main intermediate product from which other metabolic pathways can branch off. In the case of alcoholic fermentation in yeast cells, ethanol and CO_2 are produced. This purely anaerobic fermentative path (fermentation) is energetically disadvantageous for the cell compared to oxygen-required respiration, as it generates only two energy-equivalent molecules. By breaking down pyruvate to ethanol, the cell does not gain any additional ATP molecules, only those produced in glycolysis. However, it is the only way for the cell to produce energy without oxygen[12]. *Saccharomyces cerevisiae* is the yeast species primarily used here due to its tolerance for higher alcohol concentrations. You know it from the yeast blocks in the supermarket for your pizza dough.

Alcoholic fermentation is the process that we have learned to utilize well. Throughout human history, there are many historical records of the consumption of alcoholic beverages (wine in the Mesopotamian Empire[13], and date wine or beer in the Middle East[14]. Interestingly, not only we, *Homo sapiens*, are capable of enjoying and repeatedly consuming the fermentative product of alcohol, but birds have also been observed consuming alcohol through fermented berries, although this cannot be generalized[15,16]. There are also anecdotes about other animals enjoying alcohol. However, these are more stories of wild animals consuming alcoholic products from their

Basics of Fermentation

human neighbors. For instance, elephants supposedly drank South Chinese wine[17]. Stories of squirrels eating fermented pears are also found in local media. However, the squirrels did not intentionally eat the fermented pears in the garden; instead, a human had put out the pears for feeding[18]. So, even here, the consumption is not strictly voluntary.

By choosing the proper environmental conditions (oxygen: yes or no) and starting sugar concentrations, we can force yeasts to produce ethanol. The ability of microbial cells to take several pathways to gain energy (glycolysis → cellular respiration vs. fermentation) is also valid for the fermentation types mentioned later, where different end products result.

Lactic Acid Fermentation

As the name of this fermentation type suggests, the main product of the lactic acid fermentation is lactic acid. The responsible group of microorganisms are called lactic acid bacteria (LAB). However, fungi can also produce lactic acid. Biochemically, it starts again with the glycolytic splitting of hexoses and pentoses to pyruvate as an intermediate. Pyruvate is further reduced to lactic acid here. If only lactic acid results from this, we deal with obligate homofermentative mechanisms. A heterofermentative mechanism of LAB also leads to ethanol or acetic acid together with lactic acid as end products [19].

Acetic Acid Fermentation

Here again, the end product named the fermentative type, acetic acid, is produced. The previously mentioned heterofermentative mechanism depends on the starting substrate. When using pentoses (sugar molecules with five carbon atoms), only a heterofermentative metabolic pathway of LAB can be utilized, which results in acetic acid.

Additionally, a separate group of microorganisms, the acetic acid bacteria (AAB), is primarily used to produce acetic acid under strict aerobic conditions. Interestingly, AAB can directly convert the resulting ethanol into acetic acid within a multi-enzyme complex. Since these processes often occur in microbial communities, acetic acid bacteria can take up ethanol generated by other microorganisms and convert it to acetic acid [20, 21].

Basics of Fermentation

Formic Acid Fermentation

The glycolytic splitting of sugars to pyruvate is the usual metabolic pathway for fermentation to formic acid. However, there is a difference based on the capabilities of *Enterobacteria*. They have a different intracellular enzymatic setup that allows them to take paths other than those mentioned previously. Under anaerobic conditions, a metabolic shift can be induced. It leads to the selective activation of a specific enzyme (Pyruvate Formate-Lyase, the salt of formic acid is called formate), which splits pyruvate into formic acid and acetyl-coenzyme A. This shows that microorganisms can quickly adapt to changing environmental conditions to continue generating energy. It should be noted, however, that microbial formation of formic acid proceeds heterofermentatively, and further end products (ethanol, 2,3-Butanediol, acetic acid, lactic acid) are formed[22].

Homo- and Heterofermentative Microorganisms

Based on the usable end products, e.g. ethanol, acetic acid, formic acid, lactic acid, and other industrially usable organic acids, they can be divided into two larger fermentative groups. The related microorganisms produce only one desired, specific end product in the homofermentative group. In contrast, the heterofermentative group produces several usable, desired final products.

Butyric Acid Fermentation

Butyric acid fermentation also starts with the glycolytic splitting of sugar molecules (glucose, xylose, fructose). With our previous knowledge, we know the intermediate of this biological key pathway, pyruvate. Pyruvate can, as often, be further metabolized to Acetyl-Coenzyme A (Acetyl-CoA). This additional intermediate is crucial for the fermentation of butyric acid [23]. Here, again, specific bacteria are grouped under the name of butyric acid bacteria. *Clostridia* are of industrial importance, as they can produce butyric acid with a higher yield. *Clostridia* can use the previously mentioned Acetyl-CoA and convert it under strict anaerobic conditions to butyric acid using specialized enzymes. Here, a heterofermentative mechanism is happening, which is why acetic acid is simultaneously produced alongside butyric acid [24].

Acetone-Butanol-Ethanol Fermentation

Let's stay with *Clostridia* for a moment. As the name of the former pathway suggests, butyric acid is produced in butyric acid fermentation. As the reaction progresses and butyric acid production increases, the milieu surrounding the bacteria acidifies. This acidification (decreasing pH) is not particularly advantageous for the bacteria in the long run, so they switch their metabolism and move from butyric acid fermentation to acetone-butanol-ethanol fermentation. Butyric acid is reduced to butanol within an extended metabolic pathway, again heterofermentatively producing additional products [25, 26].

Basics of Fermentation

Propionic Acid Fermentation

In propionic acid fermentation, a distinct bacterial group, the *Propionibacteria*, possesses a specialized metabolic pathway. Again, glycolysis, with pyruvate as an intermediate, stands at the top of the metabolic hierarchy. However, propionic acid fermentation does not follow the classic tricarboxylic acid cycle of the cellular respiration we talked about above (the main pathway to how organic acids are produced as by-products of ATP generation). Instead, *Propionibacteria* can produce propionic acid in an additional metabolic pathway (dicarboxylic acid pathway, Wood-Werkmann fermentation) under strict anaerobic conditions [27].

Citric Acid Fermentation

Fungi are also capable of producing organic acids within their metabolic pathways. The fungus *Aspergillus niger* has made a particular name for itself in citric acid fermentation. It has been used for the industrial production of citric acid for decades. The microorganism is also brought to specific production by adjusting the environmental conditions. This is achieved through a suitable choice of energy source (sucrose) and the initial pH value, which should be below a pH of 2. This ensures a low probability of contamination with other organisms and suppresses the production of unwanted organic acids. Unlike the fermentation types mentioned before, the *A. niger*-mediated citric acid production is an aerobic process, meaning it occurs in the presence of oxygen. Often, pyruvate, coming from glycolysis, is the starting material for producing citric acid [28].

Alkaline Fermentation

At this point, we would like to introduce you to alkaline fermentation. Besides sugar molecules, other biological, energy-rich molecules can serve as substrates for fermentative processes: proteins. During the breakdown of proteins to amino acids, ammonia is released, raising the pH of the surrounding milieu to a value of 8 or higher, thus into the alkaline range. Again, this effect directly names the fermentation type. *Bacillus* species mostly carry out alkaline fermentation under aerobic conditions. Especially in African, South American, and Asian cultures, traditionally alkaline fermented foods are typical, which increase the nutritional value of the unfermented starting materials (e.g., fermented black soybeans). However, this form of fermentation plays a rather subordinate role in the processing of coffee beans [29, 30].

Now that you understand possible fermentation types, we offer a glossary of essential terms used in the coffee scene.

Basics of Fermentation

Commonly Used Terms for "Fermentation" in the Coffee Scene

In the coffee scene, several terms circulate that are only sometimes unambiguous. They often describe methods, processes, or environmental conditions.

Submerged Fermentation:
Submerged fermentation, from Latin *submersus*, "plunge under", is a fermentation that occurs surrounded by a liquid.

Anaerobic Fermentation:
It describes a fermentation without oxygen (anaerobic). The starting product is fermented with microorganisms, e.g., in a tank (bioreactor) without ventilation. Theoretically, it can be used as a step in various processing methods, whether Natural, Washed, Honey or others.

Carbonic Maceration:
Or Macération carbonique is an anaerobic fermentation. Here, whole, undamaged cherries are placed in a tank and flooded with CO_2 to displace oxygen and prevent oxidation processes. Enzymatic conversion processes occur in the cherries, and CO_2 is formed. The method originates from winemaking and can theoretically be used as a preliminary stage in all processing methods. During the microbial fermentation process (sugar to alcohol and CO_2) the coffee cherries burst due to the resulting pressure. Different fermentation conditions exist at different tank heights (from intact cherries at the top to burst cherries in submerged fermentation at the bottom) [31, 32].

Fruit Fermentation (Co-Fermentation):
When fermenting coffee, whole fruits or extracts are added to the fermentation container in order to further influence the flavor characteristics of the end product.

Semi-Carbonic Maceration (Macération Semi-Carbonique):

It describes a hybrid form of Carbonic Maceration, where a portion of intact coffee cherries is fermented with crushed cherries in a tank. No CO_2 needs to be added, as CO_2 is generated during the fermentation of the de-pulped cherries, creating the necessary protective gas atmosphere for further fermentation in the tank and displacing the oxygen upwards [33].

Koji-Fermentation:

Koji is the umbrella term for fungi of the *Aspergillus* species and is mainly associated with *Aspergillus oryzae*. The fungi is applied to the coffee beans followed by fermentation [34]. The Koji-fermented coffee is processed and dried (We even tried it ourselves).

Thermal Shock:

A temperature shock is used during fermentation to expand and quickly close the pores of the coffee cherry tissue layers (parchment and silverskin). It uses a higher temperature to penetrate flavor components into the coffee bean. Heating the fermentation tank to a higher temperature, e.g., 40 °C, is followed by a quick cold temperature drop – the shock. Sometimes, even hot and cold water is used [35]. This process is followed by the usual processing and drying, depending on the desired method.

Basics of Fermentation

Now that we have learned about the different paths of fermentation types, we will move on to the next chapter, focusing on the "actors" involved, the microorganisms. After that, we will examine how these microorganisms are utilized in the coffee processing steps.

Fig. 3.5: Ripe arabica coffee cherries of the variety Caturra, that ripen in colors red, orange and yellow in Bellavista, Peru.

References of Chapter 3

1) **Madigan MT, Martinko JM, Dunlap PV, Clark DP (2009)** Brock – Biology of Microorganisms. 12th edn. Pearson Benjamin Cummings, Munich
2) **Poltronieri P, Rossi F (2016)** Challenges in specialty coffee processing and quality assurance. Challenges 7(2):19. https://doi.org/10.3390/challe7020019
3) **Elhalis H, Cox J, Zhao J (2020)** Ecological diversity, evolution and metabolism of microbial communities in the wet fermentation of Australian coffee beans. Int J Food Microbiol 321:108544. https://doi.org/10.1016/j.ijfoodmicro.2020.108544
4) **Selmar D, Bytof G, Knopp SE et al (2006)** Germination of coffee seeds and its significance for coffee quality. Plant Biology 8(2):260–264.https://doi.org/10.1055/s-2006-923845
5) **Bytof G, Knopp SE, Schieberle P et al (2005)** Influence of processing on the generation of γ-aminobutyric acid in green coffee beans. Eur Food Res Technol 220(3-4):245–250. https://doi.org/10.1007/s00217-004-1033-z
6) **Masoud W, Bjørg Cesar L, Jespersen L et al (2004)** Yeast involved in fermentation of Coffea arabica in East Africa determined by genotyping and by direct denaturating gradient gel electrophoresis. Yeast 21(7):549–556. https://doi.org/10.1002/yea.1124
7) **Masoud W, Jespersen L (2006)** Pectin degrading enzymes in yeasts involved in fermentation of Coffea arabica in East Africa. Int J Food Microbiol 110(3):291–296. https://doi.org/10.1016/j.ijfoodmicro.2006.04.030
8) **Mathlouthi M (2001)** Water content, water activity, water structure and the stability of foodstuffs. Food Control 12(7):409–417. https://doi.org/10.1016/S0956-7135(01)00032-9
9) **Tapia MS, Alzamora SM, Chirife J (2020)** Effects of water activity (a w) on microbial stability as a hurdle in food preservation. In: Barbosa-Cánovas GV, Fontana AJ, Schmidt SJ, Labuza TP (eds) Water activity in foods. 1st edn. Wiley, New York, p 323–355

Basics of Fermentation

10) Vilela DM, de Melo Pereira GV, Silva CF et al (2010) Molecular ecology and polyphasic characterization of the microbiota associated with semi-dry processed coffee (Coffea arabica L.). Food Microbiol 27(8):1128–1135. https://doi.org/10.1016/j.fm.2010.07.024

11) Knopp S, Bytof G, Selmar D (2006) Influence of processing on the content of sugars in green Arabica coffee beans. Eur Food Res Technol 223(2):195–201. https://doi.org/10.1007/s00217-005-0172-1

12) Zamora F (2009) Biochemistry of Alcoholic Fermentation. In: Moreno-Arribas VM, Polo CM (eds) Wine Chemistry and Biochemistry. Springer, New York, p 3–26

13) Michel RH, McGovern PE, Badler VR (1993) The first wine & beer. Chemical detection of ancient fermented beverages. Anal Chem 65(8):408A–413A.https://doi.org/10.1021/ac00056a002

14) Broshi M (2007) Date Beer and Date Wine in Antiquity. Palest Explor Q 139(1):55–59. https://doi.org/10.1179/003103207x163013

15) Eriksson K, Nummi H (1983) Alcohol accumulation from ingested berries and alcohol metabolism in passerine birds. Ornis Fenn 60(1):2–9.

16) Tryjanowski P, Hetman M, Czechowski P et al (2020) Birds drinking alcohol: species and relationship with people. A Review of Information from Scientific Literature and Social Media. Animals 10(2):270. https://doi.org/10.3390/ani10020270

17) French P (2020) Fourteen elephants get drunk on corn wine and pass out in tea plantation. The Drinks Business. Union Press Ltd. Available via https://www.thedrinksbusiness.com/2020/03/fourteen-elephants-get-drunk-on-corn-wine-and-pass-out-in-tea-plantation/. Accessed 19 Feb 2024

18) The Guardian (2020) Pie-eyed and bushy-tailed: Minnesota squirrel gets drunk off fermented pears. Guardian News & Media Ltd. Available via https://www.theguardian.com/us-news/2020/nov/25/minnesota-squirrel-drunk-after-eating-fermented-pears-video. Accessed 19 Feb 2024

19) Castillo Martinez FA, Balciunas EM, Salgado JM et al (2013) Lactic acid properties, applications and production: A review. Trends Food Sci Technol 30(1):70–83. https://doi.org/10.1016/j.tifs.2012.11.007

20) Pal P, Nayak J (2016) Development and analysis of a sustainable technology in manufacturing acetic acid and whey protein from waste cheese whey. J Clean Prod 112:59–70. https://doi.org/10.1016/j.jclepro.2015.07.085

21) Gomes RJ, de Fátima Borges M, de Freitas Rosa M et al (2018) Acetic acid bacteria in the food industry: Systematics, Characteristics and Applications. Food Technol Biotechnol 56(2). DOI: 10.17113/ftb.56.02.18.5593.

22) Leonhartsberger S, Korsa I, Böck A (2002) The molecular biology of formate metabolism in enterobacteria. Journal Mol Microbiol Biotechnol 4(3):269–276

23) Zhang C, Yang H, Yang F et al (2009) Current progress on butyric acid production by fermentation. Curr Microbiol 59(6):656–663. https://doi.org/10.1007/s00284-009-9491-y

24) Jiang L, Fu H, Yang HK et al (2018) Butyric acid: Applications and recent advances in its bioproduction. Biotechnol Adv 36(8):2101–2117. https://doi.org/10.1016/j.biotechadv.2018.09.005

25) Lee SY, Park JH, Jang SH et al (2008) Fermentative butanol production by clostridia. Biotech Bioeng 101(2):209–228. https://doi.org/10.1002/bit.22003

26) Veza I, Muhamad Said MF, Latiff ZA (2021) Recent advances in butanol production by acetone-butanol-ethanol (ABE) fermentation. Biomass Bioenergy 144:105919. https://doi.org/10.1016/j.biombioe.2020.105919

27) Gonzalez-Garcia R, McCubbin T, Navone L et al (2017) Microbial propionic acid production. Fermentation 3(2):21. https://doi.org/10.3390/fermentation3020021

28) Papagianni M (2007) Advances in citric acid fermentation by Aspergillus niger: Biochemical aspects, membrane transport and modeling. Biotechnol Adv 25(3):244–263. https://doi.org/10.1016/j.biotechadv.2007.01.002

29) Owusu-Kwarteng J, Agyei D, Akabanda F et al (2022) Plant-based alkaline fermented foods as sustainable sources of nutrients and health-promoting bioactive compounds. Front Sustain Food Syst 6:885328. https://doi.org/10.3389/fsufs.2022.885328

Basics of Fermentation

30) **Samyal S (2022)** Recent trends in alkaline fermented foods. In: Singh J, Vyas A (eds) Advances in Dairy Microbial Products, Elsevier, Berlin, p 59–79
31) **Tesniere C, Flanzy C (2011)** Carbonic maceration wines. In: Advances in Food and Nutrition Research, Elsevier, Berlin, p 1–15
32) **Brioschi Junior D, Carvalho Guarçoni R, da Cássia Soares Silva M et al (2021)** Microbial fermentation affects sensorial, chemical, and microbial profile of coffee under carbonic maceration. Food Chem 342:128296. https://doi.org/10.1016/j.foodchem.2020.128296
33) **Jitjaroen W, Kongngoen R, Panjai L (2023)** Aroma profiles and cupping characteristics of coffee beans processed by semi carbonic maceration process. Coffee Sci 18:1–13. https://doi.org/10.25186/.v18i.2119
34) **Lee LW, Cheong MW, Curran P (2015)** Coffee fermentation and flavor – An intricate and delicate relationship. Food Chem 185:182–191. https://doi.org/10.1016/j.foodchem.2015.03.124
35) **Barista Hustle (2022)** Deconstructed Fermentation. Available via https://www.baristahustle.com/deconstructed-fermentation/. Accessed 14 Nov 2024

4

THE MAIN ACTORS

Chapter Summary

The key for fermentations are microorganisms, which occur naturally on coffee cherries. Thus, it is very interesting to know which microorganisms produced specific aroma precursors that formed aromas while roasting and were skillfully brewed into a cup of coffee. It is also possible to add specific enzymes or microorganisms to the fermentation vessel for a controlled and replicated fermentation. Therefore, fungi, bacteria, and cell-free enzymes are described.

A broad spectrum of naturally occurring microorganisms on coffee cherries play a role in fermentation.

Various "wild" microorganisms surround coffee cherries in their "natural state". Their presence can also be caused by other environmental factors or human intervention: for example the soil, air, water, machinery, insects, or from human skin[1-3]. Depending on the processing method and initial conditions, the populations vary in strength and diversity. Furthermore, there is even the possibility that bacteria live inside the cherry and even in the seed – these are called "endophytes." Crazy, right? Moreover, even if the pulp and mucilage are removed, fermentation could still occur inside the coffee bean[4].

Much research has been carried out to determine the exact microorganismal composition. This resulted in identifying specific advantageous starter cultures containing a defined set of microorganisms that can guide the fermentation process. Certain microorganisms have been identified that can be relevant for specific degradation processes, aroma compounds or might have negative impacts. We will get to that in a moment. First, let us introduce you to the most important bacteria and fungi. Remember that the individual microorganisms live in mixed cultures on the coffee cherry and perform taste miracles during coffee fermentation. Different types and mixed cultures of microorganisms can be present depending on environmental conditions and the method, such as Natural or Washed. And this, of course, has different influences on the aroma. For the nerds among you, a great article by Ehalis, Cox and Zhao provides a detailed overview of the diversity of organisms in the various processing methods[5]. For now, an overview will suffice as we want to start easy. ;)

The Main Actors

Fungi

Within the biological taxonomy, the classification of living beings is based on shared characteristics and is divided into animals, plants and fungi. Fungi occupy their own kingdom (they are not plants). To simplify, we will focus on the groups of yeast fungi and filamentous molds.

Yeast

Yeasts are detectable in every type of coffee processing[5]. They are usually single-celled fungi whose cells are mostly spherical or oval. For the most part, the yeasts reproduce asexually through budding or cell division (fission). They play a significant role in human life, such as in bread, beer or pizza dough production. So, we know them quite well in daily use as "degraders" of sugars and carbohydrates into carbon dioxide and ethanol. In the fermentation of coffee beans those are species such as *Saccharomyces sp., Hansinaspora sp., Candida sp.,* and *Pichia spp.*[6-8].

Yeasts can produce secondary metabolites like alcohol, acids, esters, aldehydes or ketones from the splitting of sugars in the pulp, which can be absorbed by the coffee seeds[5,6,9,10]. Thus, they significantly influence the acidity character of the coffee. You know the typical yeast flavor of a delicious piece of plum cake or a braided yeast loaf, which shows the immediate delicious influence on the pastry through yeast-mediated fermentation. An experiment in which yeasts were deliberately removed during the washed coffee processing produced a tasteless coffee[11]. So, they should not be overlooked.

Moreover, they play a role in the accelerated breakdown of the mucilage through enzyme production[1]. Some can suppress the growth of mold fungi that can produce unhealthy mycotoxins like Ochratoxin A (OTA)[12,13]. Other species *(S. cerevisiae, W. anomalus,* and *P. kudriavzevii)* possess high stability and pectinase activity even with increasing acidity [14].

Ochratoxin A

Belongs to the group of mycotoxins, toxic substances produced by fungi. Many laboratory studies have established an extensive toxicological profile for OTA. With chronic intake, OTA can exhibit carcinogenic properties, and it has also been proven to have a powerful nephrotoxic effect. Additionally, OTA can be neurotoxic and immunosuppressive. However, due to the very controversial state of research and in the spirit of the precautionary principle, the European Commission reduced the maximum content of OTA in coffee beans in May 2023[15].

Mold Fungi (Filamentous Fungi)

Mold fungi can also be found throughout the entire coffee processing chain. You're probably familiar with the "hairy look" of spoiled bread. Upon closer inspection, mold fungi consist of small, long, filamentous cells (mycelium).

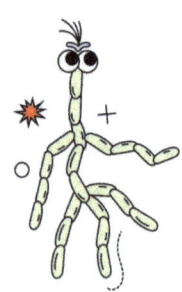

For coffee, higher numbers of mold fungus is present in the Natural and Honey processing methods than in the Washed processing[5]. They mainly occur at later stages of processing, which are drying, storage and transportation. Filamentous fungi can live well under various environ-

The Main Actors

mental conditions and are difficult to eliminate[16]. Mold fungi, such as *Aspergillus sp.* are primarily associated with adverse effects on coffee quality, e.g. by producing mycotoxins like Ochratoxin A[16]. Regarding coffee beans, this means we need to address several factors to control and prevent mold fungi to achieve good coffee quality. These include minimal soil contact during coffee processing, as few defective coffee cherries as possible[17], good ventilation, good hygiene practices and the use of GrainPro™ instead of jute for bean storage. However, a recent study showed no significant sensory differences between jute and GrainPro™[18].

Mold, as we know them from everyday life, mainly worsens the taste of coffee (moldy, musty, bad fermented). However, some species, such as *Penicillium brevicompactum*, can positively influence taste and are associated with pleasant sensory aspects like vegetal and floral[17]. Additionally, filamentous-growing Koji fungi *(Aspergillus oryzae)* are deliberately used in coffee fermentation to influence flavor profile positively[19].

GrainPro™

GrainPro™ bags are special packaging for dry agricultural products. Thanks to several gas- and moisture-impermeable plastic layers, these bags can store dry bulk goods for extended periods while preserving quality. The highly durable, reusable plastic sacks are fully recyclable. However, they increase the overall CO_2 footprint of the coffee processing chain.

Bacteria

Besides yeasts and filamentous fungi, bacteria represent the third group of microorganisms crucial for coffee fermentation. They appear in all processing methods and are helpful for a variety of commercial products, such as sauerkraut or vinegar.

Among the bacteria, lactic acid bacteria (LAB) play a significant role. This group includes *Strepto-, Entero-, Pedio-, Leuco-,* and *Lactobacilli*[5]. LABs are known for their ability to produce organic acids like glutamic acid, lactic acid, or acetic acid[20]. Other favorable products of taste quality from heterofermentative LAB are esters, aldehydes, and alcohols[21,22], but also less favorable aroma compounds such as propionic acid or butyric acid [23, 24].

In addition to lactic acid bacteria, there is a group of acetic acid bacteria (AAB). This group includes *Acetobacter spp., Gluconobacter spp.,* and *Komagataeibacter spp.*[25]. High concentrations of acetic acid can negatively impact taste with an unpleasant "vinegar taste", but low concentrations can positively influence it with a certain sweetness and a cleaner cup[26].

Some bacterial species also facilitate the breakdown of mucilage during the Washed coffee processing. These include

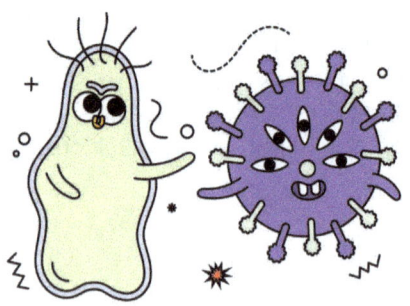

The Main Actors

Escherichia and *(Lacto-)Bacillus* species as well as *Streptococcus* and *Enterobacter* like *Klebsiella* and *Erwinia*[1,2,27]. Interestingly, during Natural coffee processing, a slightly different bacterial composition can be found, featuring, *Klebsiella, Enterobacter, Erwinia, Pseudomonas, Acinetobacter*, and *Bacillus* species [16,28,29]. During the Honey coffee processing, bacterial microorganisms are involved as well, also with a different composition of the bacterial community, mainly consisting of *Escherichia, Bacillus, Pseudomonas, Klebsiella, Enterobacter*, and *Flavobacterium* species [30,31].

Hence, there are differences in bacterial species among the various processing methods, as if the composition of the bacterial community depends on the moisture level during coffee processing. This makes sense, as bacteria are masters of adaptation, and there are adapted bacterial species for every environmental niche. Even in dry environmental conditions, bacteria have developed several survival strategies. Some spore-forming species encapsulate themselves during dry conditions and can survive as spores for a long time, or Gram-positive bacteria possess an additional protective barrier and resistance mechanism in dry conditions [32,33].

A common assumption about all bacteria is that they contribute to the breakdown of mucilage while simultaneously shaping the taste of the coffee beans through the aroma compounds they produce.

Gram-Positive and Gram-Negative Bacteria

One informal classification of bacteria is based on a staining technique by the Danish physician Hans Christian Gram (*1853, †1938). The Gram staining involves several steps using multiple dyes. Gram-positive bacteria appear blue, as they retain a blue dye within the cell wall. Gram-negative bacteria appear red, the blue dye is washed out and only the red counterstain remains.

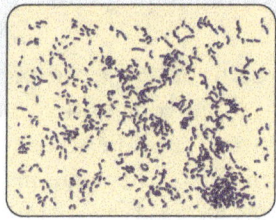 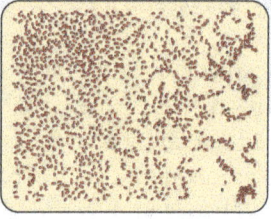

Fig. 4.1: Under the microscope, Gram-positive (blue-violet) bacteria on the left can be distinguished from Gram-negative (reddish) bacteria on the right.

Heterogeneous Cultures

As you can imagine, and we probably don't need to say much about this: An organism rarely comes alone. Just as in our macro world of animals, there is also a diverse palette of organisms among microorganisms that communally live together and depend on each other. Depending on environmental conditions, individual organisms and species can take over and prevail an environmental niche. The interplay of mixed cultures and environmental parameters significantly influences the final aroma profile of the coffee.

The Main Actors

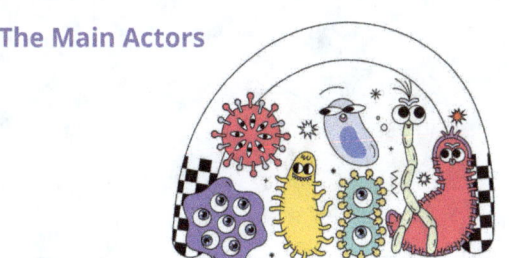

Starter Cultures

Even though organisms naturally occur on coffee cherries and are significant for fermentation, the fermentation process and the final taste experience can always vary. The random diversity of organisms and different environmental conditions ensure that fermentation processes are less controllable, efficient and consistent, and thus, the outcome can fluctuate. This "wild" fermentation can be countered by deliberately adding known cultures to the coffee cherries after harvest. You are familiar with this type of controlled fermentation from yogurt, cheese or beer[34]. There are also cultures of various microorganisms used for coffee processing. These include *Pichia*, *Saccharomyces* and *Wickerhamomyces*[35,36]. These are various mixed cultures whose selected organisms or species include targeted properties, for example, to accelerate the breakdown of mucilage. It helps to standardize the process, reduce the processing time[1,37], improve the product value, sensory quality and durability or to suppress the activity of unwanted microorganisms[13].

To be specific: For all processing methods (Natural, Honey and Washed), it has been shown that the use of unique yeast starter cultures compared to the "natural spontaneous microflora" improves and can precisely influence the taste quality of the coffee[35,38–40]. This is because selected organisms have specific properties and produce metabolic products that wild species might not deliver. Of course, this better quality only applies if the processes are well controlled and, for example, no over-fermentation occurs, which can negatively affect the taste.

Enzymes

We've been talking about microorganisms that transform substances and create new ones. If you recall, fermentation always involves enzymatically mediated transformation processes. After mentioning the use of known starter cultures to better control the fermentation process, it's also imaginable to use specific enzymes directly for the desired transformation processes instead of organisms. This is because microorganisms, especially in complex communities, inevitably lead to metabolic products that can negatively affect coffee quality as well, such as the discoloration of green coffee[1]. So why not directly use the enzymes that the bacteria would produce?

Interesting enzymes in this context include amylases, which generate simple sugars from complex sugars. Similarly, proteases can break down proteins found in mucilage-lipases, which can degrade lipids and are particularly involved in transforming aroma compounds. A large and equally important group are the pectinases. The mucilage of coffee cherry consists primarily of pectin, which these enzymes can break down. This is essential for the coffee cherry to become a coffee bean. Without pectinases, this would not be possible. Therefore, pectinases are used in the commercial sector to clarify fruit juices. They can and are used in the processing of coffee beans, but due to their relatively high price, they are used on a smaller scale and not for mass production[41]. Some companies offer products made of several enzymatic groups to suppress microbial fermentation and initiate a controlled enzymatic reaction. A few names would be Companies such as Sunson Enzymes Biotechnology LLC and Kerry Inc. (Enmex) without advertising them.

The Main Actors

References of Chapter 4

1) **Agate AD, Bhat JV (1966)** Role of pectinolytic yeasts in the degradation of mucilage layer of Coffea robusta cherries. Appl Microbiol 14 (2):256–260.
https://doi.org/10.1128/am.14.2.256-260.1966

2) **Avallone S, Guyot B, Brillouet JM et al (2001)** Microbiological and biochemical study of coffee fermentation.
Curr Microbiol 42(4):252–256. https://doi.org/10.1007/s002840110213

3) **Frank HA, Cruz ASD (1964)** Role of incidental microflora in natural decomposition of mucilage layer in Kona coffee cherries a. J Food Sci 29(6):850–853.
https://doi.org/10.1111/j.1365-2621.1964.tb00459.x

4) **Vega FE, Pava-Ripoll M, Posada F et al (2005)** Endophytic bacteria in Coffea arabica L. J Basic Microbiol 45(5):371–380.
https://doi.org/10.1002/jobm.200410551

5) **Elhalis H, Cox J, Zhao J (2023)** Coffee fermentation: Expedition from traditional to controlled process and perspectives for industrialization. Appl Food Res 3(1):100253.
https://doi.org/10.1016/j.afres.2022.100253

6) **de Melo Pereira GV, Soccol VT, Pandey A et al (2014)** Isolation, selection and evaluation of yeasts for use in fermentation of coffee beans by the wet process. Int J Food Microbiol 188:60–66.
https://doi.org/10.1016/j.ijfoodmicro.2014.07.008

7) **Poltronieri P, Rossi F (2016)** Challenges in specialty coffee processing and quality assurance. Challenges 7(2):19.
https://doi.org/10.3390/challe7020019

8) **Elhalis H, Cox J, Frank D et al (2021)** Microbiological and chemical characteristics of wet coffee fermentation inoculated with Hansinaspora uvarum and Pichia kudriavzevii and their impact on coffee sensory quality. Front Microbiol 12:713969.
https://doi.org/10.3389/fmicb.2021.713969

9) Elhalis H, Cox J, Frank D et al (2021) Microbiological and biochemical performances of six yeast species as potential starter cultures for wet fermentation of coffee beans. LWT 137:110430. https://doi.org/10.1016/j.lwt.2020.110430

10) Evangelista SR, da Cruz Pedroso Miguel MG, Silva CF et al (2015) Microbiological diversity associated with the spontaneous wet method of coffee fermentation. Int J Food Microbiol 210:102–112. https://doi.org/10.1016/j.ijfoodmicro.2015.06.008

11) Elhalis H, Cox J, Zhao J (2020) Ecological diversity, evolution and metabolism of microbial communities in the wet fermentation of Australian coffee beans. Int J Food Microbiol 321:108544. https://doi.org/10.1016/j.ijfoodmicro.2020.108544

12) Masoud W, Poll L, Jakobsen M (2005) Influence of volatile compounds produced by yeasts predominant during processing of Coffea arabica in East Africa on growth and ochratoxin A (OTA) production by Aspergillus ochraceus. Yeast 22(14):1133–1142. https://doi.org/10.1002/yea.1304

13) Haile M, Kang WH (2019) The role of microbes in coffee fermentation and their impact on coffee quality. J Food Qual 2019(1):4836709. https://doi.org/10.1155/2019/4836709

14) Haile M, Kang WH (2019) Isolation, identification, and characterization of pectinolytic yeasts for starter culture in coffee fermentation. Microorganisms 7(10):401. https://doi.org/10.3390/microorganisms7100401

15) EU-Comission (2023) Commission regulation (EU) 2023/915 of 25 April 2023 on maximum levels for certain contaminants in food and repealing Regulation (EC) No 1881/2006. Available via https://eur-lex.europa.eu/eli/reg/2023/915/oj. Accessed 18 Feb 2024.

16) Silva C, Batista L, Abreu L et al (2008) Succession of bacterial and fungal communities during natural coffee (Coffea arabica) fermentation. Food Microbiol 25(8):951–957. https://doi.org/10.1016/j.fm.2008.07.003

17) Iamanaka BT, Teixeira AA, Teixeira ARR et al (2014) The mycobiota of coffee beans and its influence on the coffee beverage. Food Res Int 62:353–358. https://doi.org/10.1016/j.foodres.2014.02.033

The Main Actors

18) **Błaszkiewicz J, Nowakowska-Bogdan E, Barabosz K et al (2023)**
    Effect of green and roasted coffee storage conditions on selected
    characteristic quality parameters. Sci Rep 13(1):6447.
    https://doi.org/10.1038/s41598-023-33609-x

19) **Rey MP (2023)** Koji processing method. Forest Coffee Fores't Blog.
    Available via
    https://coffeegreenbeans.com/blogs/producers/koji-processing-method.
    Accessed 14 Feb 2024

20) **Zalán Z, Hudáček J, Štětina J et al (2010)** Production of organic acids
    by Lactobacillus strains in three different media.
    Eur Food Res Technol 230(3):395–404.
    https://doi.org/10.1007/s00217-009-1179-9

21) **Ribeiro LS, da Cruz Pedrozo Miguel MG, Evangelista SR et al (2017)**
    Behavior of yeast inoculated during semi-dry coffee fermentation and
    the effect on chemical and sensorial properties of the final beverage.
    Food Res Int 92:26–32.
    https://doi.org/10.1016/j.foodres.2016.12.011

22) **Zhang SJ, de Bruyn F, Pothakos V et al (2019)** Following coffee
    production from cherries to cup: Microbiological and metabolomic
    analysis of wet processing of Coffea arabica. Appl Environ Microbiol
    85(6):e02635-18.
    https://doi.org/10.1128/AEM.02635-18

23) **Amorim HV, Amorim VL (1977)** Coffee enzymes and coffee quality.
    In: Robert LO, Allen JSA (ed) Enzymes in food and beverage
    processing, vol 47. ACS Symposium Series, Washington D.C., p 27–56

24) **Goto YB, Fukanaga ET (1986)** Coffee: Harvesting and processing for
    top quality coffee. Honolulu (HI): Hawaii Agricultural Experiment
    Station, University of Hawaii

25) **Gomes RJ, de Fátima Borges M, de Freitas Rosa M et al (2018)**
    Acetic acid bacteria in the food industry: Systematics, Characteristics
    and Applications. Food Technol Biotechnol 56(2).
    DOI: 10.17113/ftb.56.02.18.5593

26) **Bertrand B, Boulanger R, Dussert S et al (2012)** Climatic factors directly impact the volatile organic compound fingerprint in green Arabica coffee bean as well as coffee beverage quality.
Food Chem 135(4):2575–2583.
https://doi.org/10.1016/j.foodchem.2012.06.060

27) **Avallone S, Brillouet JM, Guyot B et al (2002)** Involvement of pectolytic micro-organisms in coffee fermentation.
Int J Food Sci Tech 37(2):191–198.
https://doi.org/10.1046/j.1365-2621.2002.00556.x

28) **Silva CF, Schwan RF, Sousa Dias Ë et al (2000)** Microbial diversity during maturation and natural processing of coffee cherries of Coffea arabica in Brazil. Int J Food Microbiol 60:(2-3):251–260.
https://doi.org/10.1016/S0168-1605(00)00315-9

29) **Silva CF, Vilela DM, de Souza Cordeiro C et al (2013)** Evaluation of a potential starter culture for enhance quality of coffee fermentation. World J Microbiol Biotechnol 29(2):235–247.
https://doi.org/10.1007/s11274-012-1175-2

30) **Velmourougane K (2013)** Impact of natural fermentation on physicochemical, microbiological and cup quality characteristics of Arabica and Robusta coffee.
Proc Natl Acad Sci, India, Sect B Biol Sci 83(2):233–239.
https://doi.org/10.1007/s40011-012-0130-1

31) **Vilela DM, de M Pereira GV, Silva CF (2010)** Molecular ecology and polyphasic characterization of the microbiota associated with semi-dry processed coffee (Coffea arabica L.). Food Microbiol 27(8):1128–1135.
https://doi.org/10.1016/j.fm.2010.07.024

32) **Hirai Y (1991)** Survival of bacteria under dry conditions; from a viewpoint of nosocomial infection. J Hosp Infect 19(3):191–200.
https://doi.org/10.1016/0195-6701(91)90223-U

33) **Qiu Y, Zhou Y, Chang Y et al (2022)** The effects of ventilation, humidity, and temperature on bacterial growth and bacterial genera distribution. IJERPH 19(22):15345.
https://doi.org/10.3390/ijerph192215345

The Main Actors

34) Laranjo M (2023) Starter culture and their role in fermented foods. Christon JH (ed) Microbial fermentations in nature and as designed processes. 1st edn. Wiley, New York p 281–292
35) Evangelista SR, da Cruz Pedrozo Miguel MG, de Souza Cordeiro C et al (2014) Inoculation of starter cultures in a semi-dry coffee (Coffea arabica) fermentation process. Food Microbiol 44:87–95. https://doi.org/10.1016/j.fm.2014.05.013
36) Mahingsapun R, Tantayotai P, Panyachanakul T et al (2022) Enhancement of Arabica coffee quality with selected potential microbial starter culture under controlled fermentation in wet process. Food Biosci 48:101819. https://doi.org/10.1016/j.fbio.2022.101819
37) Massawe GA, Lifa SJ (2010) Yeasts and lactic acid bacteria coffee fermentation starter cultures. IJPTI 2(1):41. https://doi.org/10.1504/IJPTI.2010.038187
38) de Melo Pereira GV, Neto E, Soccol VT et al (2015) Conducting starter culture-controlled fermentations of coffee beans during on-farm wet processing: Growth, metabolic analyses and sensorial effects. Food Res Int 75:348–356. https://doi.org/10.1016/j.foodres.2015.06.027
39) de Melo Pereira GV, Beux M, Pagnoncelli MGB et al (2016) Isolation, selection and evaluation of antagonistic yeasts and lactic acid bacteria against ochratoxigenic fungus Aspergillus westerdijkiae on coffee beans. Lett Appl Microbiol 62(1):96–101. https://doi.org/10.1111/lam.12520
40) Evangelista SR, Silva CF, da Cruz Miguel MGP et al (2014) Improvement of coffee beverage quality by using selected yeasts strains during the fermentation in dry process. Food Res Int 61:183–195. https://doi.org/10.1016/j.foodres.2013.11.033
41) Kashyap DR, Vohra PK, Chopra S et al (2001) Applications of pectinases in the commercial sector: a review. Bioresour Technol 77(3):215–227. https://doi.org/10.1016/S0960-8524(00)00118-8

5

COFFEE PROCESSING

Chapter Summary

The most common primary types of post-harvest coffee processing are Natural, Washed or hybrid methods like Honey. Coffee processing aims to remove the layers surrounding the coffee seeds and dry the green coffee beans to a stable residual moisture level. This is crucial for the storability, overall quality and sensory experience of the cup. Different processing methods produce various sensory qualities in coffee. The main reason for the variety of methods are the environmental conditions at the farms and the existing traditions. Today, processing methods are also being experimented with in places where this was traditionally not common.

Fig.5.1: Ripe (red) and unripe (green) coffee cherries on the same branch.

Coffee Processing

Let's delve into the methods of coffee processing and their differences, keeping all the details in mind as we eventually explore the possibilities of fermentation in coffee processing and its impact on coffee flavor.

The most common primary types of post-harvest coffee processing are Natural, Washed or hybrid methods like Honey. Coffee processing aims to remove the layers surrounding the coffee seeds (coffee beans) – the pulp, mucilage, parchment, and silver skin – and to dry the green coffee beans to a stable residual moisture level[1]. Balanced drying is crucial for the storability, overall quality and sensory experience of the cup[2].

Utilization of Coffee Mucilage

All processing methods involve the removal of the mucilage, which is the slimy layer between the pulp and the parchment skin. This layer is naturally vital for the fermentative step within coffee processing. After removing the coffee cherry, the mucilage can be used for further bioprocesses. It is mainly composted. Successful field studies have also been conducted for the production of (bio-)hydrogen from the remaining coffee mucilage and other biological residues.[3-5]. Another approach is the production of bioethanol. Here, successful experiments have also been carried out. From bioethanol production it is not a far leap to a targeted fermentative production of alcoholic beverages from various coffee by-products: the mucilage or pulp. Successful laboratory experiments have also been conducted to produce alcoholic beverages, yielding promising results. The authors note that further optimization must be done to produce a flavorful and high-quality alcoholic beverage[6]. Further uses of other coffee by-products are well presented in a review article by Cássia Campos[7].

Why are There so Many Different Methods?

You have already seen that different processing methods produce different sensory qualities in coffee. A primary reason for the variety of methods are the environmental conditions at the cultivation sites and the associated traditions. Today, various cultivation and processing methods are also being experimented with in places where this was traditionally not common. However, there is a limiting effect: the climate. The choice of processing method is not only determined by the demand for specific flavor profiles but by two crucial environmental factors: humidity and water availability. They mainly decide the processing and drying conditions[2]. Natural fermentation is possible in a dry climate, sun drying succeeds easily and mold formation is suppressed without high humidity.

However, if the humidity is high, mold formation is risky. Therefore, other methods are necessary to remove the pulp and dry the coffee seeds in humid climates. Those methods use for example de-pulpers and water for the first steps to get rid of material that might mold. Later those coffees dry under roofs, in greenhouses, or with machines. Here's a graphical representation. You see several methods divided according to weather, humidity and water conditions.

Coffee Processing

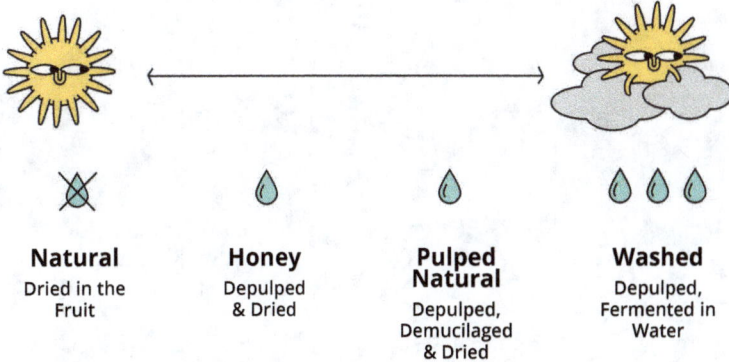

Fig. 5.2: Depicted are regional climate characteristics, the water usage and the final coffee processing methods.

Natural / Dry-Processed / Sun-Dried

The harvested ripe fruits are pre-sorted and then spread out on the ground (soil, concrete, asphalt, or on tarps), on drying tables, or in solar dryers/greenhouses in a layer that is not too thick for drying and regularly turned (there are also drying machines). At night, the fruits are piled up and covered. Within 10 – 25 days, the cherries dry in the sun. During this time, fermentation processes occur[8,9]. The organisms release enzymes that break down the pulp and mucilage, allowing the contained moisture to evaporate more easily[10]. You are already familiar with this.

Fig. 5.3: Natural processing of coffee. The cherries are spread on a raised bed in a drying tent in Marcala, Honduras.

Fig. 5.4: Dried Natural-processed coffee after twenty days on a raised airy drying patio in Marcala, Honduras.

Solar Dryer

Utilizes the higher air temperature generated by solar energy in a greenhouse to dry coffee beans faster[9]. Compared to sun drying, it accelerates the process by up to 50%[11]. Moreover, the greenhouse covers the coffee from environmental influences like weather, insects, additional microorganisms, or mycotoxins. It proves to be helpful in rainy regions to dry large quantities of coffee quickly, which produces coffee with higher quality, lower investment and energy costs compared to artificial drying methods[9].

Fig. 5.5: Solar dryer for parchment coffee on the farm of Maria Gimenez, Bellavista, Peru.

Coffee Processing

Constant mixing is essential to reduce moisture evenly and prevent defects such as over-fermentation, unwanted decomposition, or mold formation[2,9]. After drying to a residual moisture below 12%, the pulp is removed and the coffee beans are stored for several weeks to improve shelf life[12]. The parchment is removed shortly before shipping and the beans are sorted and packaged. Fermentation times are longer for Naturals than for other methods[2]. In terms of taste, they present a complex number of aroma-active building blocks responsible for the presence of a silky mouthfeel, complex aromas and wild fruitiness. However, the brewed coffees tend to taste less clear.

Fig. 5.6: Stored parchment coffee at a collection centre of the Coop Cenfrocafe in Jaén Peru waiting for transport to a drymill for export preparation.

Fig. 5.7: Export preparation – sieving and sorting of the coffee to exclude coarse particles at the drymill of Cenfrocafe in Peru.

Fig. 5.8: Process of coffee sorting regarding bean density and size, Cenfrocafe, Peru.

Fig. 5.9: Colour sorting process to enhance coffee-quality, Cenfrocafe, Peru.

Fig. 5.10: Visual quality difference of sorted coffee on the left and the excluded beans on the right.

Drying and Taste

There are different mechanical dryers used to dry coffee quickly in a reproducible manner without losing aroma. The used temperatures are below 50 °C to retain the aromatic precursors. Studies show good results in comparison to sun dried coffee when using energy efficient heat pump dryers that enable accurate control of the temperature and relative humidity of the drying gas. The drying process resulted in a very similar chemical composition compared to sun dried coffee[13] a study with robusta showed a superior preservation of aroma precursors[14]. There are also mechanical dryers which are coupled to a low carbon dioxide or nitrogen atmosphere. As the beans aren't exposed to oxygen, volatile compounds could be retained even better within the final product[15].

Washed

First, the harvested cherries are placed into a water tank to remove the floating unripe cherries (also known as floaters). Then, the remaining cherries are de-pulped, meaning they are squeezed between two adjustable rollers, separating them from most of the peel and pulp. Afterward, the remaining pulp ferments in water tanks for 1 to 3 days[8,16]. As you already know, microorganisms help dissolve the thick solid layer of pectin[10]. Stopped at the right moment (no one likes over-fermentation), the coffee is dried in one of the previously described ways and regularly turned. Unlike dry processing, this wet fermentation consumes considerable amounts of water, depending on the regional way of processing and the reuse of water. In Washed coffee processing 20 to 90 Liter of water is used to wash one kilogram of coffee cherries[17]. This wastewater really needs decontamination to prevent groundwater pollution, but sadly it is not a common thing. We will dive into that in chapter 7.

Coffee Processing

Wet/Washed processing allows for better control of process parameters, resulting in coffee generally having fewer defects and better quality than Natural processing[18].

Fig. 5.11: Typical water tank for Washed coffee and a depulper.

Fig. 5.12: Inside view of a depulper with a sieve for size.

Fig. 5.13: Sun drying of washed parchment coffee at Cenfrocafe in Jaén, agitated by the coffee producer Roman Flores Segundo Ariol.

Pulped Natural / Honey / Semi-Dry

The Pulped Natural process from Brazil closely resembles the Central American Honey process. In both, defective cherries are sorted out first within a water tank as before (floaters)[2]. The cherry enclosed by the pulp is mechanically de-pulped.

Pulped Natural

This method uses knives and centrifuges to remove the pulp. The green coffee is dried after mechanical mucilage removal and washing steps. The remaining mucilage layer dries on the cherries, contributing to the sensory experience later in the cup. Good Pulped Naturals thus have high sweetness, balanced acidity, and a creamy mouthfeel. Due to the rapid removal of mucilage, the least fermentation occurs compared to other methods[2].

Honeys / Semi-Dry

The amount of pulp that remains on the coffee seed is adjusted using a de-pulper. The amount of remaining pulp on the coffee seed is distinguished by terms from "yellow" for little and "red" to "black" for a lot of pulp. The "sticky" coffee seeds are dried and fermented on tarps on the ground or drying tables[19]. The amount of remaining pulp determines the duration and intensity of the fermentation.

Sun, cloudiness, or shade influence the duration of drying. Depending on how much pulp was left on the cherry, the moisture content of the de-pulped coffee seeds can be determined. As we already know, this is a critical fermentation parameter.

Fig. 5.14: Depulping process – Cherries are transferred in a hopper (funnel) and the depulped parchment beans are collected in a plastic bag for fermentation.

Fig. 5.15: Honey processing of coffee. The sticky parchment beans are spread out by the producer Medim Jimenez on airy patios under a roof.

Fig. 5.16: Dried Honey processed coffee – Clearly visible is the brownish dried pulp and mucilage layer on the yellow parchment.

Although the coffee, like in the Natural process, must be turned daily, the risk of defect formation is significantly lower[2]. The remaining pulp after de-pulping allows for a greater body (silky mouthfeel) and sweetness compared to Washed coffees[20].

Burundian

This method is called a "double soak" or "double fermentation". The wet, depulped coffee seeds are fermented in water for 24 hours before a wash removes "floaters". The remaining seeds are then fermented in water a second time. Afterward, the coffee is dried[21]. This process influences the remaining and infused metabolites of the beans through fermentation. According to Kwon and colleagues, it results in more sugars and organic acids and less caffeine, asparagine, gamma-aminobutyric acid, and chlorogenic acid[22].

Semi-Washed / "Giling Basah" / Wet-Hulled

This method is widespread in Indonesia; the coffee is spread out after depulping but only pre-dried to 25 – 35% moisture content instead of 11 – 12% while being turned. The coffee seeds are then removed from the parchment shell (which in other methods happens just before shipping) and dried as green coffee until it reaches a safe residual moisture content for storage and shipping. This process allows coffee farmers to dry the coffee faster, even under unfavorable high humidity conditions, reducing the time from picking to shipping to about a month. In comparison, Central American coffee takes 2 – 3 months. This method typically produces earthy, woody, herbal, nutty and roasted flavor nuances, as well as low acidity and a full body[21,23].

Animal Digestive Fermentation

As we aim to comprehensively describe the various fermentation processes of coffee, we have to mention this method. However, from an ethical standpoint, we do not welcome it and urge you to consider other coffee styles instead. Indeed, this process can already be artificially replicated if the taste is preferred[24]. Some things should be left alone by humans and this definitely includes using animals as machines and most abhorrent, keeping them in cages to "refine" coffee.

Animal fermentations include, for example, Kopi Luwak (cat-like mammals), Black Ivory from elephants (*Elephas maximus*), or coffee from the Jacu Bird. Here, coffee consumed by the animals is collected from the feces after the digestive process, cleaned, sufficiently dried, removed from the parchment shell and sorted. There is also Monkey Parchment coffee, where monkeys chew the pulp of ripe coffee cherries and spit out the remaining parchment cherries like cherry pit spitting by us[25-27]. The food consumed with coffee cherries, for example, tropical fruits, plays a role in developing the coffee's flavor profile[28]. The fermentation in the digestive tract lasts 12 hours for Kopi Luwak and 12 to 70 hours for elephants[25,26]. For Kopi Luwak, it was found that the taste difference is significantly caused by protein degradation and enrichment with free amino acids[28]. The amino acid valine contributes to the taste, body and balance. The formation of glutamic acid and histamine influences the flavor, aroma and acidity. Kopi Luwak is also said to have less bitterness, which can be due to the formation of shorter amino acids and peptides due to lower initial proteins through degradation processes (proteolysis)[29].

Coffee Processing

Monsooned

This method, widespread in India, replicates an accidentally created process from the times of sailing ships, when coffee from the British Indian colonies was shipped to Europe. The transported coffee was exposed to the humidity of the monsoon rains during the voyage and continued to mature during transport. The coffee becomes milder, loses acidity and acquires specific spicy flavor nuances. Today, the process is deliberately replicated due to the ongoing demand for its taste. For this purpose, coffee processed as Natural is exposed to the monsoon rains in warehouses and then dried[30]. During the so-called monsooning, changes in the coffee beans' microbial population occur. Here, *Aspergillus* species are noteworthy, which dominate after approximately four weeks of monsooning[31].

"Modern Techniques"

Chapter 3 introduced different fermentations, including those common in the coffee scene under names like "anaerobic" or "carbonic maceration". These are types of fermentations that are theoretically applicable to all methods. Thus, specific anaerobic fermentation steps can be performed in barrels or bags at various points in the process. Therefore, they are anaerobically fermented, in addition to Washed and Naturally processed coffees. Now that you know the basic processing methods, you probably already know when certain fermentations occur or which microorganisms might play a role. In the next chapter, we will combine all this knowledge and finally focus on the influence on taste. Ooh yeah!

References of Chapter 5

1) **Silva CF, Vilela DM, de Souza Cordeiro C et al (2013)** Evaluation of a potential starter culture for enhance quality of coffee fermentation. World J Microbiol Biotechnol 29(2):235–247.
https://doi.org/10.1007/s11274-012-1175-2

2) **Poltronieri P, Rossi F (2016)** Challenges in specialty coffee processing and quality assurance. Challenges 7(2):19.
https://doi.org/10.3390/challe7020019

3) **Hernández MA, Rodríguez Susa M, Andres Y (2014)** Use of coffee mucilage as a new substrate for hydrogen production in anaerobic co-digestion with swine manure. Bioresour Technol 168:112–118.
https://doi.org/10.1016/j.biortech.2014.02.101

4) **Cárdenas E, Zapata-Zapata A, Kim D (2019)** Hydrogen production from coffee mucilage in dark fermentation with organic wastes. Energies 12(1):71. https://doi.org/10.3390/en12010071

5) **Rangel CJ, Hernández MA, Mosquera JD et al (2021)** Hydrogen production by dark fermentation process from pig manure, cocoa mucilage, and coffee mucilage. Biomass Conv Bioref 11(2):241–250.
https://doi.org/10.1007/s13399-020-00618-z

6) **Kc Y, Subba R, Shiwakoti LD et al (2021)** Utilizing coffee pulp and mucilage for producing alcohol-based beverage. Fermentation 7(2):53.
https://doi.org/10.3390/fermentation7020053

7) **Campos RC, Pinto VRA, Melo LF et al (2021)** New sustainable perspectives for "Coffee Wastewater" and other by-products: A critical review. Future Foods 4:100058.
https://doi.org/10.1016/j.fufo.2021.100058

8) **Schwan RF, Wheals AE (2003)** Mixed microbial fermentations of chocolate and coffee. In: Boekhout T, Robert V. (ed) Yeasts in Food. Woodhead Publishing Ltd, Cambridge, p 429–449.
https://doi.org/10.1533/9781845698485.429

9) **Firdissa E, Mohammed A, Berecha G et al (2022)** Coffee drying and processing method influence quality of Arabica coffee varieties (Coffee arabica L.) at Gomma I and Limmu Kossa, southwest Ethiopia. J Food Qual 2022(1):9184374.
https://doi.org/10.1155/2022/9184374
10) **Silva CF, Schwan RF, Sousa Dias Ë et al (2000)** Microbial diversity during maturation and natural processing of coffee cherries of Coffea arabica in Brazil. Int J Food Microbiol 60(2-3):251–260.
https://doi.org/10.1016/S0168-1605(00)00315-9
11) **Esper A, Mühlbauer W (1998)** Solar drying - an effective means of food preservation. Renew Energ 15(1–4):95–100.
https://doi.org/10.1016/S0960-1481(98)00143-8
12) **Silva C, Batista L, Abreu L et al (2008)** Succession of bacterial and fungal communities during natural coffee (Coffea arabica) fermentation. Food Microbiol 25(8):951–957.
https://doi.org/10.1016/j.fm.2008.07.003
13) **Kulapichitr F, Borompichaichartkul C, Suppavorasatit I et al (2019)** Impact of drying process on chemical composition and key aroma components of Arabica coffee. Food Chem 291:49–58.
https://doi.org/10.1016/j.foodchem.2019.03.152
14) **Dong W, Hu R, Long Y et al (2019)** Comparative evaluation of the volatile profiles and taste properties of roasted coffee beans as affected by drying method and detected by electronic nose, electronic tongue, and HS-SPME-GC-MS. Food Chem 272:723–731.
https://doi.org/10.1016/j.foodchem.2018.08.068
15) **Barista Hustle (2022)** Deconstructed Fermentation. Available via https://www.baristahustle.com/deconstructed-fermentation/. Accessed 14 Nvv 2024
16) **Fowler MS, Leheup P, Cordier JL (1998)** Cocoa, coffee and tea. In: Wood BJB (ed) Microbiology of Fermented Foods. Springer, Boston, p 128–147

17) **Ulsido MD, Geleto MZ, Berego YS (2024)** Waste water management in wet coffee processing mills and their impact on the water quality status of Gidabo river and its tributaries, southern Ethiopia. Environ Health Insights 18:1–11. https://doi.org/10.1177/11786302241260953

18) **Lee LW, Cheong MW, Curran P et al (2015)** Coffee fermentation and flavor – An intricate and delicate relationship. Food Chem 185:182–191. https://doi.org/10.1016/j.foodchem.2015.03.124

19) **Vilela DM, de M Pereira GV, Silva CF et al (2010)** Molecular ecology and polyphasic characterization of the microbiota associated with semi-dry processed coffee (Coffea arabica L.). Food Microbiol 27(8):1128–1135. https://doi.org/10.1016/j.fm.2010.07.024

20) **Knopp S, Bytof G, Selmar D (2006)** Influence of processing on the content of sugars in green Arabica coffee beans. Eur Food Res Technol 223(2):195–201. https://doi.org/10.1007/s00217-005-0172-1

21) **Febrianto NA, Zhu F (2023)** Coffee bean processing: Emerging methods and their effects on chemical, biological and sensory properties. Food Chem 412:135489. https://doi.org/10.1016/j.foodchem.2023.135489

22) **Kwon DJ, Jeong HJ, Moon H et al (2015)** Assessment of green coffee bean metabolites dependent on coffee quality using a 1H NMR-based metabolomics approach. Food Res Int 67:175–182. https://doi.org/10.1016/j.foodres.2014.11.010

23) **Sweet Maria's Coffee Inc. (2021)** Sumatran Coffee Processing: Why You Should Know Giling Basah. Available via https://library.sweetmarias.com/why-should-you-know-giling-basah/. Accessed 15 Dec 2024

24) **Sudarti, Bektiarso S, Prastowo SHB et al (2020)** Optimizing lactobacillus growth in the fermentation process of artificial civet coffee using extremely- low frequency (ELF) magnetic field. J. Phys.: Conf. Ser. 1465(1):12010. DOI: 10.1088/1742-6596/1465/1/012010.

25) **Raveendran A, Murthy PS (2022)** New trends in specialty coffees - "the digested coffees". Crit Rev Food Sci Nutr 62(17):4622–4628. https://doi.org/10.1080/10408398.2021.1877111

26) **Thammarat P, Kulsing C, Wongravee K et al (2018)** Identification of volatile compounds and selection of discriminant markers for elephant dung coffee using static headspace gas chromatography—mass spectrometry and chemometrics. Molecules 23(8):1910. https://doi.org/10.3390/molecules23081910
27) **Haile M, Bae HM, Kang WH (2020)** Comparison of the antioxidant activities and volatile compounds of coffee beans obtained using digestive bio-processing (Elephant Dung Coffee) and commonly known processing methods. Antioxidants 9(5):408. https://doi.org/10.3390/antiox9050408
28) **Patria A, Abubakar A, Febriani MM (2018)** Physicochemical and sensory characteristics of luwak coffee from Bener Meriah, Aceh-Indonesia. IOP Conf Ser: Earth Environ Sc. 196:12010. DOI: 10.1088/1755-1315/196/1/012010.
29) **Marcone MF (2004)** Composition and properties of Indonesian palm civet coffee (Kopi Luwak) and Ethiopian civet coffee. Food Res Int 37(9):901–912. https://doi.org/10.1016/j.foodres.2004.05.008
30) **Ahmad R, Tharappan B, Bongirwar DR (2003)** Impact of gamma irradiation on the monsooning of coffee beans. J Stored Prod Res 39(2):149–157. https://doi.org/10.1016/S0022-474X(01)00043-1
31) **Tharappan B, Ahmad R (2006)** Fungal colonization and biochemical changes in coffee beans undergoing monsooning. Food Chem 94(2):247–252. https://doi.org/10.1016/j.foodchem.2004.11.016

6

LET'S PUT IT ALL TOGETHER –
FROM FERMENTATION TO CUP

Chapter Summary

Specific fermentation steps can be performed at various points during coffee processing. The fermentation itself requires several elements to succeed: Coffee cherries, microorganisms, environmental conditions and the processing methods. Depending on the technique, a different amount of cherry-biomass remains on the parchment during processing and drying. Additionally, the varied use of water and other parameters like temperature, pH, ripeness, sugar content, microorganisms, duration of fermentation, altitude or weather are key influencing factors. This results in specific flavor developments depending on the method used and depending on the microorganisms and fermentation steps used within the process.

Fig. 6.1: A bucket full of fresh picked coffee cherries.

Let's put it all Together

It's time – you are now biotech-experts without too much prior knowledge and can participate in discussions about fermented coffees. Ultimately, fermentation requires several elements to come together and you are now familiar with the respective variables:

Coffee cherries, microorganisms, environmental conditions and the respective processing methods. Depending on the technique, a different amount of cherry-biomass remains on the parchment during processing and drying. Additionally, the varied use of water (none vs. a lot of water, washing steps and the duration in water) and other parameters like temperature, pH, ripeness, sugar content, microorganisms, duration of fermentation, altitude or weather also play a role.

You can surely imagine that environmental conditions have a significant impact. At what altitude is fermentation carried out? What are the daytime and nighttime temperatures? What is the humidity? Producers or processing stations must naturally experiment with fermentations, depending on the climate and location. This also means that fermentation is not a standardized process that can easily be transferred in its nuances. This results in very specific local practices and knowledge. Interesting, right?

In this chapter, you can apply your newly acquired knowledge and answer how fermentation affects coffee's taste.

We have created a graphic that tries to give a comprehensive overview of the different fermentation possibilities. Don't worry – we will explain it.

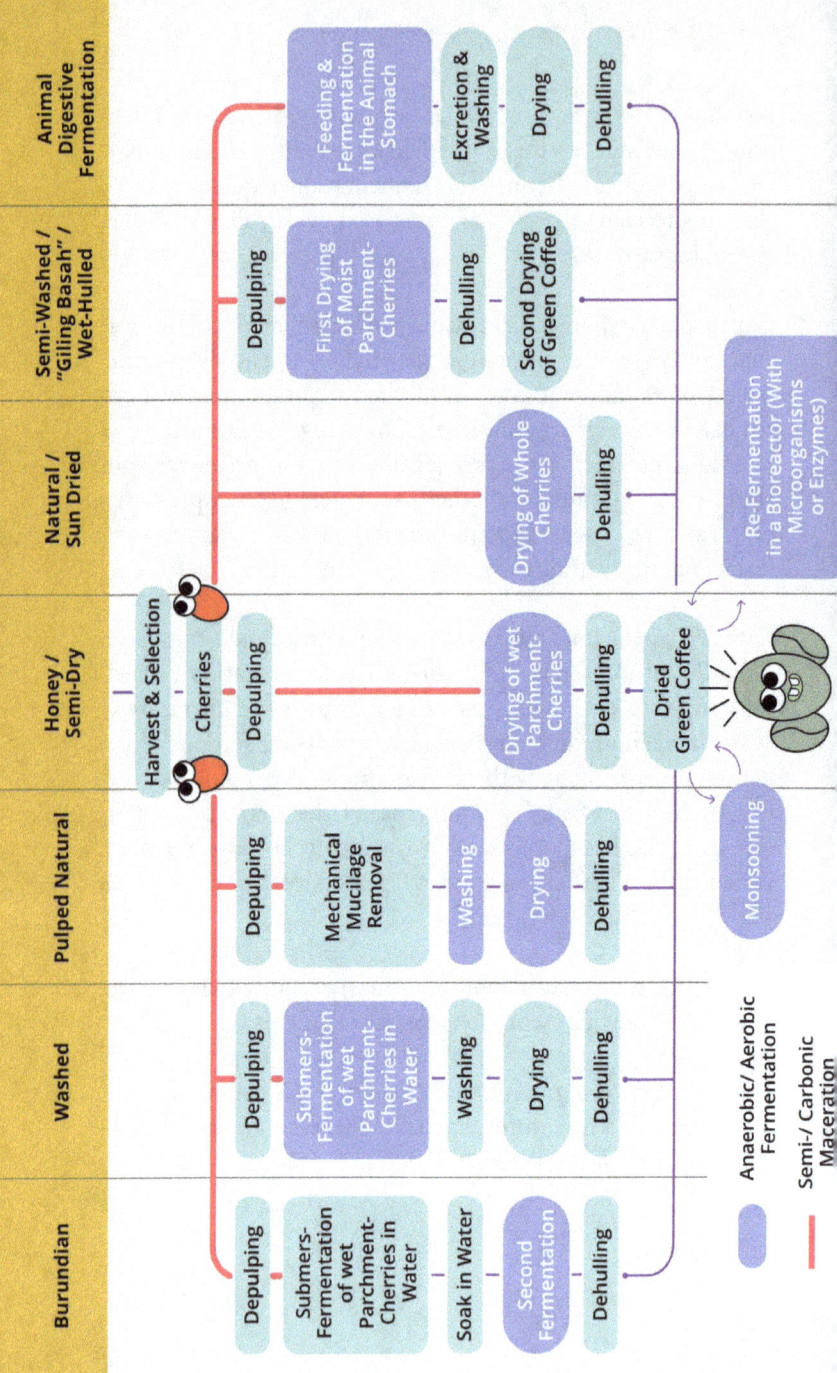

Let's put it all Together

Previous Page
Fig. 6.2: Categorized coffee processing steps. Purple boxes and red lines mark possible steps to introduce fermentation.

- In the various columns, you will find the most common methods for processing coffee, such as Natural, Washed, Honey, etc.
- The individual coffee processing steps are described, from the cherries to the processed green coffee beans (raw coffee). We have already introduced the details of each method in the previous chapter.
- Marked in purple and red, you will find the steps involving fermentation. Here, it would be best to recall the various possible "Environmental influencing factors in the fermentation process" from chapter 3, that affect the taste pathways: initial conditions, environmental conditions such as the temperature, competition, etc.

Fermentation and its Impact on Sensory Perception

1. Negative Influences by Over- or Under-Fermentation.

Control of the process parameters is essential to avoid taste and aroma errors. Under-fermentation can lead to the unwanted proliferation of microorganisms, such as filamentous fungi, which contribute to the formation of undesirable substances due to the reduced breakdown of mucilage[1,2]. Over-fermentation creates an environment for organisms that increases the amounts of acetic acid, propionic acid and butyric acid, which can also negatively affect quality, e. g. by creating an onion-like taste [2-4].

2. Selected Starter Cultures and Their Influence on Coffee Sensory Perception.

Remember that various starter cultures can produce certain fundamental aroma compounds. We want to present some of them here, without being complete (Table 6.1). Since the relationships are very complex, they can only serve as examples for the respective process. However, we still want to present what has been discovered in research.

Microorganisms	Fermentation	Sensory Effects
C. parapsilosis and *S. cerevisiae* (Yeast)	Natural / Honey	caramel and fruity, favorable taste [5]
Pichia fermentans (Yeast)	Washed	intense vanilla and floral aroma [6]
Torulaspora delbrueckii (Yeast)	Natural Anaerob	sweet, yet acidic, citric and caramell [7,8]
Kefir (heterogeneous culture of yeasts and bacteria)	Re-Fermentation	fruity, nutty, acidic, caramel and spicy, which depends on the fermentation time [9]
Lactiplantibacillus plantarum (Bacteria)	Washed	fruity with slightly acidic notes [10,11]
Leuconostoc mesenteroides (Bacteria)	Washed	citric and herbal components [11]

Tab. 6.1: Selected microorganisms as starter cultures and their influence on coffee aroma.

3. Anaerobic Fermentation / Semi- / Carbonic Maceration of Whole/ Pulped/ Partially Pulped Cherries in Bioreactors, Tanks, or Plastic Bags

This fermentation can be a preliminary stage for all processing methods – whether Natural, Washed or Honey. Pereira and colleagues describe that anaerobic fermentation enhances the activity of lactic acid bacteria and yeasts, thereby increasing the production of essential aroma compounds[12]. Compared to other methods, a different composition of organic components emerges: more chlorogenic acid, sucrose, lactic acid, trigonelline and less glucose. Sensory-wise, this results in more fermented/ tangy and sour notes.

Here are two research examples, of course, that only represent individual examples.
In the first one, an anaerobic-Natural fermented coffee resulted in a flavor profile described as floral, fruity, winey, woody, caramel and alcoholic.
While the second an anaerobic-Washed processed coffee resulted in a flavor profile of floral and fruity. With the addition of citrus and sugarcane flavor notes [12,13].
Depending on the choice of starter cultures, additional effects are possible. With *S. cerevisiae*, for example, enhanced aromas of citrus fruit, floral, plum or peach emerge compared to anaerobic fermentation without starter cultures[14].

Interestingly, anaerobic fermentation can produce many aroma components and significantly impact roasted coffee's antioxidant properties[15]. So, drinking coffee could be even healthier than previously thought ;)

Fig. 6.3: Sunset in the region of Chota, Peru.

4. Fermentation in Water Tanks During Washing

Fermentation in water tanks also enhances coffee's aromatic and sensory properties by increasing important foundational components for forming aromatic compounds during roasting: organic acids such as aspartic or glutamic acid, esters, alcohols and aldehydes. This also includes lowering the pH[16,17]. Moreover, submerged fermentation conveniently reduces the concentration of undesirable compounds like butyric acid, negatively affecting taste[17,18]. Washed coffees are known for their zesty acidity and clear, floral, caramel and fruity taste notes[19].

Let's put it all Together

Fig. 6.4: Dried Washed-processed coffee in its yellow parchment, stored in bags before transportation to the drymill.

5. Fermentation During Drying

Naturally processed coffees have a longer fermentation time, a more complex composition of aroma-active substances, more body (silky mouthfeel), less acidity and a broader range of fruit notes, often resulting in funky fruitiness. Depending on the method, Pulped Naturals and Honeys have less acidity and more body than Washed coffees[20]. This is because the remaining pulp after depulping allows for more body and sweetness compared to Washed coffees[21]. And now you say, of course! Because we have more fermentation material for our microorganisms. We have aerobic conditions since the coffee is outdoors and not underwater like in the Washed process. Therefore, less sugar is broken down, and more reducing sugars like glucose and fructose remain[21].

Bytof and colleagues showed that during drying, metabolic processes within the coffee seeds also influence the amount and composition of the amino acid content[16] – you remember: amino acids are also essential building blocks for the aromas later during roasting. This happens depending on the drying duration and thus, the respective method, as Washed coffees dry faster than Naturally processed ones.

6. Fermentation During Storage

Since the coffee seeds are dried to a stable residual moisture content at which essentially no biological activity occurs anymore, fermentation plays a minor role during storage. Only unsanitary practices, e.g. due to increased moisture during transport or insufficient drying beforehand, could promote mold formation, which can be responsible for undesirable aromas or the formation of OTA[22]. Taste changes due to long storage are attributed to the excessive aging of green coffee beans[23] and not fermentation.

7. Animal Fermentation and "Imitated Animal Fermentation"

Coffee fermented by animals is characterized by lower acidity, less body and a richer taste of tropical fruits[24,25]. It isn't easy to make general statements about the taste when imitating this process in bioreactors. Due to the challenge of replicating the microbial diversity of animal fermentation, the outcome varies depending on the choice of microorganisms and the raw coffee. For instance, the fermentation of raw coffee with yeasts (S*accharomyces cerevisiae* and *Pichia kluyveri*) reduces acidity and sweetness. Still, it enhances caramel, nut and roast aromas in the cup[26]. Or, coffee fermented with enzymes from *Aspergillus oryzae* is characterized by smoothness and favorable taste, mouthfeel and aftertaste[27].

8. Fermentation of Green Coffee Beans in a Bioreactor /Re-Fermentation /Solid-State Fermentation

It is also possible to ferment already-dried green coffee beans with starter cultures or enzymes. Thus, it is not a fermentation in the classical sense during processing up to drying. This method can always be applied retrospectively and plays a role, especially in research[2,26]. To what extent this aroma and taste modulation will find broad commercial application remains to be seen. It may become more relevant with the advancing anthropogenic climate change, to sensorially improve coffee on a large scale or to compensate for fluctuating coffee quality. We can tell from our own experiments with re-fermentation, that *Aspergillus oryzae* (Koji) and *Rhizopus oligosporus* (Tempeh) can indeed (positively) influence the flavor.

Clever minds have written an algorithm for machine learning to keep track of all fermentation types, especially the combination possibilities. With further developments, finding the optimal fermentation and processing combinations may make achieving the desired taste on the tongue easier [28].

References of Chapter 6

1) **Lee LW, Cheong MW, Curran P (2015)** Coffee fermentation and flavor – An intricate and delicate relationship. Food Chem 185:182–191. https://doi.org/10.1016/j.foodchem.2015.03.124
2) **Haile M, Kang WH (2019)** The role of microbes in coffee fermentation and their impact on coffee quality. J Food Qual 2019(1):4836709. https://doi.org/10.1155/2019/4836709
3) **Amorim HV, Amorim VL (1977)** Coffee enzymes and coffee quality. In: Robert LO, Allen JSA (ed) Enzymes in food and beverage processing, vol 47. ACS Symposium Series, Washington D.C., p 27–56
4) **Osorio Pérez V, Álvarez-Barreto CI, Matallana LG et al (2022)** Effect of prolonged fermentations of coffee mucilage with different stages of maturity on the quality and chemical composition of the bean. Fermentation 8(10):519. https://doi.org/10.3390/fermentation8100519
5) **Evangelista SR, da Cruz Pedrozo Miguel MG, de Souza Cordeiro C Silva et al (2014):** Inoculation of starter cultures in a semi-dry coffee (Coffea arabica) fermentation process. Food Microbiol 44:87–95. https://doi.org/10.1016/j.fm.2014.05.013
6) **de Melo Pereira GV, Neto E, Soccol VT et al (2015)** Conducting starter culture-controlled fermentations of coffee beans during on-farm wet processing: Growth, metabolic analyses and sensorial effects. Food Res Int 75:348–356. https://doi.org/10.1016/j.foodres.2015.06.027
7) **da Mota MCB, Batista NN, Rabelo MHS et al (2020)** Influence of fermentation conditions on the sensorial quality of coffee inoculated with yeast. Food Res Int 136:109482. https://doi.org/10.1016/j.foodres.2020.109482
8) **Bressani APP, Martinez SJ, Sarmento ABI et al (2020)** Organic acids produced during fermentation and sensory perception in specialty coffee using yeast starter culture. Food Res Int 128:108773. https://doi.org/10.1016/j.foodres.2019.108773

9) Afriliana A, Pratiwi D, Giyarto et al (2019) Volatile compounds changes in unfermented robusta coffee by re-fermentation using commercial kefir. NFSIJ 8(4). DOI: 10.19080/NFSIJ.2019.08.555745
10) de Carvalho Neto DP, de Melo Pereira GV, Finco AMO et al (2018) Efficient coffee beans mucilage layer removal using lactic acid fermentation in a stirred-tank bioreactor: Kinetic, metabolic and sensorial studies. Food Biosci 26:80–87. https://doi.org/10.1016/j.fbio.2018.10.005
11) Ribeiro LS, da Cruz Pedrozo Miguel MG, Martinez SJ et al (2020) The use of mesophilic and lactic acid bacteria strains as starter cultures for improvement of coffee beans wet fermentation. World J Microbiol Biotechnol 36(12):186. https://doi.org/10.1007/s11274-020-02963-7
12) Pereira TS, Batista NN, Santos Pimenta LP (2022) Self-induced anaerobiosis coffee fermentation: Impact on microbial communities, chemical composition and sensory quality of coffee. Food Microbiol 103:103962. https://doi.org/10.1016/j.fm.2021.103962
13) Martins PMM, Batista NN, da Cruz Pedroz Miguel MG et al (2020) Coffee growing altitude influences the microbiota, chemical compounds and the quality of fermented coffees. Food Res Int 129:108872. https://doi.org/10.1016/j.foodres.2019.108872
14) Martinez SJ, Rabelo MHS, Bressani APP et al (2021) Novel stainless steel tanks enhances coffee fermentation quality. Food Res Int 139:109921. https://doi.org/10.1016/j.foodres.2020.109921
15) Várady M, Tauchen J, Klouček P et al (2022) Effects of total dissolved solids, extraction yield, grinding, and method of preparation on antioxidant activity in fermented specialty coffee. Fermentation 8(8):375. https://doi.org/10.3390/fermentation8080375
16) Bytof G, Knopp SE, Schieberle P et al (2005) Influence of processing on the generation of γ-aminobutyric acid in green coffee beans. Eur Food Res Technol 220(3-4):245–250. https://doi.org/10.1007/s00217-004-1033-z
17) Elhalis H, Cox J, Frank D et al (2020) The crucial role of yeasts in the wet fermentation of coffee beans and quality. Int J Food Microbiol 333:108796. https://doi.org/10.1016/j.ijfoodmicro.2020.108796

18) Elhalis H, Cox J, Frank D et al (2021) The role of wet fermentation in enhancing coffee flavor, aroma and sensory quality. Eur Food Res Technol 247(2):485–498. https://doi.org/10.1007/s00217-020-03641-6
19) Gonzalez-Rios O, Suarez-Quiroz ML, Boulanger R et al (2007) Impact of "ecological" post-harvest processing on the volatile fraction of coffee beans: I. Green coffee. J Food Compost Anal 20(3-4):289–296. https://doi.org/10.1016/j.jfca.2006.07.009
20) Poltronieri P, Rossi F (2016) Challenges in specialty coffee processing and quality assurance. Challenges 7(2):19. https://doi.org/10.3390/challe7020019
21) Knopp S, Bytof G, Selmar D (2006) Influence of processing on the content of sugars in green Arabica coffee beans. Eur Food Res Technol 223(2):195–201. https://doi.org/10.1007/s00217-005-0172-1
22) Silva C, Batista L, Abreu L et al (2008) Succession of bacterial and fungal communities during natural coffee (Coffea arabica) fermentation. Food Microbiol 25(8):951–957. https://doi.org/10.1016/j.fm.2008.07.003
23) Selmar D, Bytof G, Knopp SE (2008) The storage of green coffee (Coffea arabica): Decrease of viability and changes of potential aroma precursors. Ann Bot 101(1):31–38. https://doi.org/10.1093/aob/mcm315
24) Marcone MF (2004) Composition and properties of Indonesian palm civet coffee (Kopi Luwak) and Ethiopian civet coffee. Food Res Int 37(9):901–912. https://doi.org/10.1016/j.foodres.2004.05.008
25) Noel MG, Lebrilla CB, Garcia EV (2016) A Multi-Omics Approach to Finding Biomarkers in Philippine Civet Coffee. Proc DLSU Arts Congress 1(4)
26) Wang C, Sun J, Lassabliere B et al (2020) Coffee flavour modification through controlled fermentation of green coffee beans by Saccharomyces cerevisiae and Pichia kluyveri: Part II. Mixed cultures with or without lactic acid bacteria. Food Res Int 136:109452. https://doi.org/10.1016/j.foodres.2020.109452

27) **Murthy PS, P Sneha H, Basavaraj K et al (2019)** Modulation of coffee flavor precursors by Aspergillus oryzae serine carboxypeptidases. LWT 113108312.
https://doi.org/10.1016/j.lwt.2019.108312
28) **Rocha RAR, da Cruz MAD, Silva LCF et al (2024)** Evaluation of Arabica Coffee fermentation using machine learning. Foods 13(3):454.
https://doi.org/10.3390/foods13030454

7

SUSTAINABILITY & WASTEWATER

Chapter Summary

The processing of coffee has environmental impacts especially when water is used. Microorganisms and other substances of the coffee cherry pollute the used water. This used wastewater can lead to the pollution of whole ecosystems, rivers and groundwater and pose lasting health risks for humans. In the following chapter, causes and effects are discussed and examples for good practice are described. If you want to be more sustainable, then it is better to obtain sun dried Naturals that use less water in the overall process.

Over the past decades, one thing has become clear: global coffee consumption increases continually, and with it, there is a need for a sustainable cultivation and processing strategy, especially in our time of continuing climate catastrophe. Regarding coffee producers, climate and pest adaptation is in full swing. New varieties, for example, by crossing arabicas with robustas, will bring possible climatically adapted plants and new flavor profiles.

Despite all the romance about microorganisms, there's one more thing. The microbial contamination of water used for coffee processing can lead to the pollution of ecosystems, rivers and groundwater. This is a big underestimated problem, especially with Washed coffee, as many fincas and washing stations do not treat their wastewater enough. A recent study in Peru showed that 85% of small coffee producers do not treat their wastewater at all[1].

After depulping coffee cherries, the wastewater-sludge is contaminated with microorganisms, a mixture of sometimes toxic substances, such as nitrate, phosphate, caffeine, phenols, melanoidins, lignins and tannins. It is acidic (has a low pH around 4), has a high biological oxygen demand and contains suspended matter[2-4].

Phew, we can tell you: this is quite a lot of contamination, and this untreated wastewater can pollute surrounding ecosystems and water bodies downstream and pose health risks for aquatic life and humans. Proper treatment of those waters is not a common thing. Studies, for example in Ethiopia, revealed those negative effects and made clear that the treatment performance has to be improved as national and international threshold values have not been met yet[5,6]. Therefore, the control of wastewater treatment must continue to be a focus of sustainable fermentation practices in coffee production.

Fig. 7.1: A depulper at work, separating the parchment cherries from their red to yellow skin and pulp. The skin and pulp will be composted and later used for fertilizing the coffee trees.

The standard wastewater treatment procedure uses different approaches depending on the socio-economic situation. Here are some examples. An option in Ethiopia could be the usage of big constructed wetlands that benefit from higher climatic temperatures for biodegradation in a controlled manner[6].

This is possible if there is no land scarcity and land is available for reasonable prices[7]. Possible is also the combination of mechanical and biochemical processes. So, Enden and Calvert suggest several steps before the introduction into these wetlands: An acidification pond to reduce the sedimentable solids, limestone treatment of the clean effluents to raise the pH from 4 to 6, and followed by fermentation in an anaerobic bioreactor[2]. Yes, fermentation is used again, but this time, the processes help clean up the wastewater through aerobic and anaerobic fermentation. Fermentation is just simply amazing, right? The wetlands use several steps of plants and microorganisms to reduce the amount of phosphate and organic matter by aerobic fermentation and, afterward, a pond with water hyacinths to reduce bacteria and heavy metals. Even using sieves to reduce the number of pectins and cellulose fibers would be an easy option to reduce the environmental impact[8]. Also, the use of Vetiver, a bunchgrass, has proven its effectiveness in several countries[1]. Using those plants as a filter is comparatively cheap and needs no strict maintenance if the wastewater is cleaned of organic matter that could clog the system.

Some examples of implementing those practices with further benefits for the farms and the environment are e.g. the use of ion-exchange technologies in wastewater treatment[9] or the production of biogas, which can be produced by fermenting coffee residuals[10].

From an "I want to be environmentally friendly" point of view, Natural coffees are the stars in this respect. They use less water during coffee processing and have a lower environmental impact, as less wastewater is produced. Sure, other aspects play a role too, when considering the whole environmental impact, such as plantation type, organic farming, transport and so forth.

Sustainability & Wastewater

References of Chapter 7

1) **MOCCA (2023):** Vetiver: la iniciativa de bajo costo para tratar las aguas mieles. Available via https://mocca.org/vetiver-la-iniciativa-de-bajo-costo-para-tratar-las-aguas-mieles/. Accessed 27.10.2024

2) **Enden J, Calvert KC (2010)** Review of coffee wastewater characteristics and approaches to treatment. Available via https://www.researchgate.net/publication/238084098_Review_of_coffee_wastewater_characteristics_and_approaches_to_treatment. Accessed 27 Oct 2024

3) **González-Freire A, Martínez-Hernández CM (2022)** Mejoramiento y uso de los efluentes para el beneficio del café. Revista Ciencias Técnicas Agropecuarias 31(2)

4) **Genanaw W, Kanno GG, Derese D et al (2021)** Effect of wast water discharge from coffee processing plant on river water quality, Sidama region, south Ethiopia. Environ Health Insights 15. https://doi.org/10.1177/11786302211061047

5) **Beyene A, Kassahun Y, Addis T et al (2012)** The impact of traditional coffee processing on river water quality in Ethiopia and the urgency of adopting sound environmental practices. Environ Monit Asseess 184:7053–7063. https://doi.org/10.1007/s10661-011-2479-7

6) **Ulsido MD, Geleto MZ, Berego YS (2024)** Waste water management in wet coffee processing mills and their impact on the water quality status of Gidabo river and its tributaries, southern Ethiopia. Environ Health Insights 18:1–11. https://doi.org/10.1177/11786302241260953

7) **Alemayehu YA, Asfaw SL, Tirfie TA (2019)** Management options for coffee processing wastewater. A review. J Mater Cycles Waste 22(2):454–469. https://doi.org/10.1007/s10163-019-00953-y

8) **Keppeler T, Nadolski L, Romero C (2023)** Kaffee: Eine Geschichte von Genuss und Gewalt, 2nd edn. Rotpunktverlag, Zurich, p 46

9) **Ijanu EM, Kamaruddin MA, Norashiddin FA (2020)** Coffee processing wastewater treatment: a critical review on current treatment technologies with a proposed alternative. Appl Water Sci 10(1). https://doi.org/10.1007/s13201-019-1091-9
10) **Rodríguez-Valencia N, Franco DAZ, Ramírez Gómez CA (2013)** Manejo y disposición de los subproductos y de las aguas residuales del beneficio del café – En Federación Nacional de Cafeteros de Colombia. Cenicafé Manual del cafetero colombiano: Investigación y tecnología para la sostenibilidad de la caficultura 3:111–136

8

SOME LAST WORDS

You may feel the same as us: fermentation is quite exciting, and we wanted to show that it has always been significant for the flavor development of coffee, not just a rising trend among coffee nerds. What is new, however, is the deliberate experimentation with various methods and the profound knowledge about microorganisms and their effects on the sensory qualities of one of the most important beverages of modern times. Imitating animal-fermented coffee may hopefully lead to the end of the ethically questionable practice of this kind of fermentation. Ultimately, knowledge about new flavor variants can lead to sustainable practices (e.g., using less water).

We hope that the world of coffee fermentation becomes more familiar to you through our microscope. A lot is happening in research and practice right now – which is why it wouldn't surprise us, if there were new findings and further methods for coffee fermentation soon, especially as the demand for fermented coffee continues to grow. However, the basics of the processes remain the same – only the taste will change. :)

Fig. 8.1: A cupping of microlots at the coffee lab of the cooperative Rutas del Inca in Querocoto, Peru.

The Coffee of the Future?

As a small excursion into current basic research, we have something special for you: coffee plant cells cultivated in the laboratory. They are comparable to the "lab-grown meat" frequently mentioned in the news. Specifically, these cells were isolated from coffee leaves and cultured in a Petri dish. The cells were subsequently washed and roasted. A comparative tasting using commercial, roasted coffee varieties showed a taste similarity in roast aromas and bitterness. However, comparable caffeine concentrations could not be achieved. So, until lab coffee shows a similarly invigorating effect, we must remain patient and wait for further research projects[1].

Some Last Words

References of Chapter 8

1) Aisala H, Kärkkäinen E, Jokinen I et al (2023) Proof of concept for cell culture-based coffee. J Agric Food Chem 71(47):18478–18488. https://doi.org/10.1021/acs.jafc.3c04503

Further Thoughts of us

Apart from focusing on the exquisite taste of coffees, sustainability, and fair wages within the coffee production chain are also essential to us. Yes, political responsibility has a significant impact, but it is not the only thing that masters. We, as consumers, can have a direct effect and push markets in a different direction. One such thing should be to buy organic and fair trade coffee instead of conventional, mass-produced coffee. And to combine those aspects in balance with good taste, go for organic & fair specialty coffee. Further, you could support your local roasters and specifically ask for coffee directly marketed by farmers or cooperatives. Thereby, you can ensure that farmers are paid fairly. This comes with an increased price and might not be available for everyone who enjoys coffee. However, small steps matter and are needed to induce a more considerable shift.

Fig. 8.2: Vinzenz Särchen

Fig. 8.3: Marcel Hackler

© The Author(s), under exclusive license to Springer Nature Switzerland AG 2025
M. Hackler, V. Särchen, *The Art and Science of Coffee Fermentation*, https://doi.org/10.1007/978-3-031-91599-4

Acknowledgement

During the creation, we received much support from our immediate environment. We want to extend our special thanks to: Fred Armbrust, Achim Barghorn, Yasmin Cordova, Kira Datschun, Nina Gmeiner, Hermine Hackler, Oliver Klitsch, Simone König, Ina Kudlich, Philipp Mertens, Nikolaus & Cordula Särchen, Carolin Selig, Christopher Simon, and Philipp Zingler. Your content suggestions and critical comments have greatly improved the comprehensibility and readability of this book. We also thank the Käthe Kaffeerösterei GmbH, Flying Roasters and Oatly for their great support. We thank Philipp Tresbach for his passionate commitment to provide us with impressive photos. And finally we thank our two graphic designers: Sarah Diedrich thanks for your creative ideas and patience throughout finalizing this project. And a big hug goes to Hander Cadavid Parra, you motivated us throughout our writing process and the moments when we wanted to quit with your humorous figures and icons. Many thanks to all of you for believing in this project.

The illustrations in the book were made by Hander Cadavid Parra.

Thanks for your financial support:

© The Author(s), under exclusive license to Springer Nature Switzerland AG 2025
M. Hackler, V. Särchen, *The Art and Science of Coffee Fermentation*, https://doi.org/10.1007/978-3-031-91599-4

Glossary

A

AAB – Acetic Acid Bacteria.

Acetyl-CoA (Acetyl Coenzyme A) – The most important intermediate product in the cell metabolism of the three main nutrients carbohydrates, lipids, and amino acids.

Aerobic/Anaerobic – The term comes from the ancient greek word 'aer' – meaning air. It refers to the condition under which an organism metabolizes.

Anaerobic Fermentation – Fermentation in the absence of oxygen.

Animal Fermentation – Relies on the digestion and thus fermentation of coffee cherries by animals.

Alkaline/Basic – Measure of the basic character of an aqueous solution, which has a high pH value.

Aspergillus Niger – A globally occurring black mold

ATP – Adenosine triphosphate – the cell's energy carrier.

B

Bacillus – A group of rod-shaped bacterial single-cells that can survive under aerobic as well as anaerobic conditions.

Bacteria – Single-celled organisms without a nucleus.

Biocatalytic – Transformation and acceleration of chemical reactions (catalysis) in which enzymes serve as biological catalysts.

Biochemical – Describes chemical processes in organisms.

Biological Systematics – Classification of living beings into different groups along phylogenetic relationships.

Burundian – A method of coffee fermentation that relies on double fermentation in water.

Butanol – Alcohol with four carbon atoms.

C

Caffeine – A psycho-stimulating chemical compound, originally obtained from plants.

Carbohydrate – One of the most important biomolecules alongside proteins and fats. It is composed of carbon, oxygen, and hydrogen. The simplest form is simple sugar.

Carbonic Maceration – Fermentation in the absence of oxygen by using CO_2.

Cellular Respiration – A metabolic process in which energy is obtained from the oxidation of organic substances.

CGA – Chlorogenic Acid, an ester product of caffeic acid and quinic acid, has an important antioxidant effect.

Citric Acid Cycle – A biochemical process in the metabolism of aerobically growing biological cells.

Clostridia – A group of bacterial single-cells that are ubiquitous and can only survive under anaerobic conditions (absence of oxygen).

Coffee Cherry – The stone fruit of the coffee plant, which contains seeds, colloquially called coffee beans.

E

Enterobacteria – A group of bacteria primarily found in the intestines of humans and animals.

Enzyme – A protein that can accelerate chemical reactions.

Ethanol – Commonly drinkable alcohol, volatile, highly flammable, and colorless.

F

Fatty Acid – A long-chain hydrocarbon chain equipped with a chemically functional acid group that results in an acidic environment in aqueous surroundings. There are saturated and unsaturated types, depending on the presence of chemically reactive double bonds.

Fermentation – An enzymatic metabolic process in which organic substances are converted. Organic acids, alcohols, and gases are produced. It serves the biological cell to produce energy.

Flesh – Tissue in fruits with a high water content that surrounds the seeds, see also pulp.

Fruit Fermentation (Co-Fermentation) – Whole fruits or extracts are added to the fermentation container in order to further influence the flavor characteristics of the end product.

Fungi – Organisms that do not perform photosynthesis, require 'food' as an energy source, live relatively sessile and are the ecosystem's main decomposer.

G

Germination – The beginning of the development of the seed, the first stage in the emergence of a new plant.

Glycolysis – The breakdown of sugars in living beings.

GrainPro^{TM} – Plastic bags used in addition to jute bags for the storage and transport of raw coffee.

H

Hexose – A simple sugar with a carbon skeleton of six carbon atoms.

Homo- and Heterofermentative – Indicates whether one or multiple products are produced during fermentation.

Honey – Coffee processing where the crushed cherries are dried.

Hydrogel – Gel that can bind water due to its large surface area and three-dimensional network structure.

I

Intermediate Product – Product of a chemical reaction that is further processed.

K

Koji Fermentation – Uses fungi from the Koji family for fermentation. These play a crucial role in Japanese products like miso, soy sauce, or sake.

L

LAB – Lactic Acid Bacteria.

Lipids – A group name for water-insoluble substances like fatty acids or wax.

M

Maillard Reaction – A non-enzymatic browning reaction where aroma compounds are reformed. Known from roasting and frying.

Metabolic Processes – Also called metabolism, refers to all chemical transformations of substances in the body of living beings. Includes anabolism and catabolism.

Metabolites – Intermediate product on a metabolic pathway.

Methylmalonyl-CoA – An important intermediate product in the use of fatty acids as an energy source.

Microbial – Of, from, by microorganisms.

Microorganisms – Microscopically small organisms, including but not limited to bacteria, fungi, and algae, primarily single-celled.

Molds – Belong to the group of filamentous fungi and are known as fluffy colored coating on spoiled foods.

Monsooned – Coffee processing where the raw coffee is exposed to monsoon rains in warehouses.

Mucilage – Colorless viscous pectin layer, mucous layer.

N

Natural – Coffee processing where the whole cherries are dried.

O

Organic Acids – Carbon-based chemical substances that have several functional groups, which can donate a proton in an aqueous solution and acidify the surrounding environment.

Organism – Biological term for a living being in its entirety of cells.

OTA – Ochratoxin A, a mold toxin produced by many fungi.

P

Parchment Skin – Intermediate layer in the coffee cherry that surrounds the coffee seeds.

Pectins – Plant carbohydrates, known as dietary fiber and plant gelling agent.

Pentoses – Simple sugars with a carbon skeleton of five carbon atoms.

pH Value – Indicates the measure of the acidic or basic character of an aqueous solution.

Proteins – Biological macromolecule composed of amino acids, colloquially called protein.

Pulp – The term for the flesh of coffee.

Pulped Natural – Coffee processing where the crushed cherries are further processed and dried.

Pyruvate – An important intermediate product in aerobic and anaerobic metabolism.

R

Reducing Sugars – Biochemically important molecules that can have a reducing effect due to a chemically functional group. They can donate electrons, reduce other substances, and are themselves oxidized.

S

Semi-carbonic Maceration – A hybrid form of carbonic maceration, where a part of intact coffee cherries is fermented with already crushed cherries in a tank. CO_2 is produced in the process.

Semi-dry – See Honey.

Semi-washed – Coffee processing where the green coffee seeds are immediately mechanically removed from all fruit layers of the coffee cherry and directly dried.

Starter Culture – A known mixture of microorganisms with specific properties for targeted fermentation.

Submerged Fermentation – Fermentation within a liquid.

T

Thermal Shock – A temperature shock is used during the fermentation process to expand and afterwards quickly close the pores of the coffee cherry tissue layers to bring in and lock up flavor components.

W

Washed – Coffee processing where the fermentable fruit layers of the crushed cherries are removed by submerged fermentation in water.

Y

Yeasts – Single-celled fungi, e.g., known from the kitchen.

GPSR Compliance

The European Union's (EU) General Product Safety Regulation (GPSR) is a set of rules that requires consumer products to be safe and our obligations to ensure this.

If you have any concerns about our products, you can contact us on

ProductSafety@springernature.com

In case Publisher is established outside the EU, the EU authorized representative is:

Springer Nature Customer Service Center GmbH
Europaplatz 3
69115 Heidelberg, Germany

www.ingramcontent.com/pod-product-compliance
Lightning Source LLC
LaVergne TN
LVHW010344260326
834688LV00036B/865